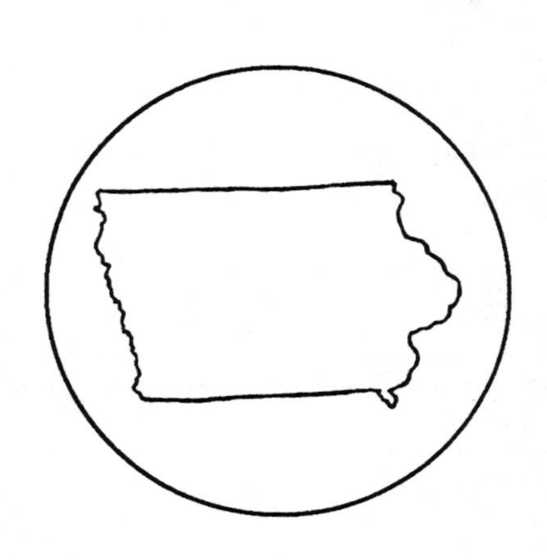

IOWA

ENVIRONMENTAL LAW HANDBOOK

Editors

Steven M. Christenson and William J. Keppel

of

Dorsey & Whitney

801 Grand Avenue, Suite 3900
Des Moines, IA 50309
515-283-1000

1330 Connecticut Avenue, N.W.
Washington, D.C. 20036
202-452-6900

220 South Sixth Street
Minneapolis, MN 55402
612-340-2600

Contributors

B. Andrew Brown
Rebecca A. Comstock
Gregory A. Fontaine
Mark R. Kaster
Thomas S. Llewellyn
George Y. Sugiyama
Robert E. Cattanach
William A. Dossett
Elizabeth A. Goodman
Alexandra B. Klass
Kevin J. Saville

Government Institutes, Inc.

Government Institutes, Inc.
4 Research Place, Suite 200
Rockville, Maryland 20850

Published January 1995

02 01 00 5 4 3

ISBN: 0-86587-429-9

Library of Congress Catalog Card Number: 94-74196

Printed in the United States of America.

IOWA ENVIRONMENTAL LAW HANDBOOK

SUMMARY OF CONTENTS

IOWA ENVIRONMENTAL LAW HANDBOOK

TABLE OF CONTENTS

Chapter One: Agencies With Responsibility for Environmental Programs

Chapter Two: Air Pollution Control

Chapter Three: Water Pollution Control

Chapter Four: Wetlands and Protected Waters

Chapter Five: Superfund: Hazardous Conditions Cleanup

Chapter Nine: Solid Waste Management

Chapter Ten: Storage Tank Regulation

Chapter Eleven: Agricultural Chemicals

Chapter Twelve: Biotechnology: Genetically Engineered Organisms

Chapter Thirteen: Asbestos, PCB, and Lead Regulation

Chapter Sixteen: Environmental Considerations in Real Estate and Business Transactions

Chapter Seventeen: Pollution Prevention and Compliance Auditing

PREFACE

The Iowa Environmental Law Handbook is intended to provide general guidance to the complex statutes and regulations that comprise environmental law in Iowa. We have attempted to include Iowa statutes and rules adopted before November 1, 1994, and federal statutes and regulations where Iowa agencies enforce them or where Iowa law is incomplete on a topic. From this body of environmental law, we have emphasized requirements that we believe will be of greatest interest to the Iowa business community with special attention to certain agricultural and municipal issues.

Although this handbook aims to be as comprehensive and accurate as possible given the space constraints, it is merely an overview and is not intended as a substitute for legal advice. Environmental laws often are not models of clarity, and legal interpretations usually depend on the facts and circumstances of specific cases. Moreover, environmental laws change frequently and environmental regulations interpreting those laws are continuously being promulgated, amended, and updated. In addition, court decisions may expand or limit the application of these laws and regulations. Accordingly, the reader should obtain the assistance of a knowledgeable environmental attorney when confronted with environmental issues and problems.

Several acknowledgments are in order. We thank the attorneys in Dorsey & Whitney's Environmental and Regulatory Affairs Department for their contributions to this handbook. In addition, we thank Dorsey summer associates Peter Beckerman, William Bird, Ingo Burghardt, Julie Gyurci, and Richard Murphy for their assistance in editing and checking the accuracy of the thousands of citations in the footnootes. We also thank numerous staff members of the Iowa Department of Natural Resources who kindly reviewed various draft chapters and offered helpful suggestions, although mistakes that remain are ours. Thanks are also due to Susan Trapskin who typed this handbook with care and dedication. Finally, we thank Dorsey & Whitney for its support in making this handbook possible.

Steven M. Christenson
William J. Keppel
January 1, 1995

ABOUT THE AUTHORS

All of the authors are attorneys with the law firm of Dorsey & Whitney.

Steven M. Christenson. Born 1963, Spencer, Iowa. Admitted to bar: Minnesota, 1988; Colorado, 1989; U.S. District Court, District of Minnesota, 1989; U.S. Court of Appeals, 8th Circuit, 1989. *Education*: Iowa State University (B.A., 1985); University of Iowa (J.D., 1988). Mr. Christenson clerked for the Honorable George G. Fagg, U.S. Court of Appeals, 8th Circuit, 1988-1989. *Publications*: Minnesota Environmental Law Handbook, co-editor, (2d ed. 1994); The Superfund Act: An Overview, co-author, 4 Canada-U.S. Bus. L. Rev. 147 (1990); Regulatory Jurisdiction over non-Indian Hazardous Waste in Indian Country, 72 Iowa L. Rev. 1091 (1987). (612-340-5603)

William J. Keppel. Born 1941, Sheboygan, Wisconsin. Admitted to bar: Minnesota, 1970; U.S. District Court, District of Minnesota, 1970; U.S. Supreme Court, 1979. *Education*: Marquette University (A.B., 1963); University of Wisconsin (J.D., 1970). Mr. Keppel is a former law professor. *Publications include*: Minnesota Environmental Law Handbook, co-editor, (2d ed. 1994); Minnesota Civil Practice (4 vols), co-author, (2d ed. 1990); Minnesota Administrative Practice and Procedure, co-author (1982). (612-340-2745)

B. Andrew Brown. Born 1957. Admitted to bar: District of Columbia, 1987; Minnesota, 1990. *Education:* Stanford University (A.B., 1979); Harvard University (M.P.A., 1981); Duke University (J.D., 1986). From 1986 until 1990, Mr. Brown practiced at Donovan, Leisure, Newton & Irvine and Willkie, Farr & Gallagher in Washington, D.C. (612-340-5612)

Robert E. Cattanach, Jr. Born 1947. Admitted to bar: Wisconsin, 1975; U.S. Supreme Court, 1980; Minnesota, 1983. *Education:* United States Naval Academy (B.S., 1972); University of Wisconsin (J.D., 1975). *Publications*: Handbook of Environmentally Conscious Manufacturing (1994); Encyclopedia of Environmental Clauses and Forms (expected publication, 1995). (612-340-2873)

Rebecca A. Comstock. Born 1950. Admitted to bar: Minnesota, 1978. *Education:* University of Minnesota (B.A., 1973); University of Denver (J.D., 1977). *Publications:* The Superfund Act: An Overview, co-author, 4 Canada-U.S. Bus. L. Rev. 147 (1990); Wetlands Regulation in Minnesota, 46 Bench & Bar, No. 3 (Mar. 1989). (612-340-2987).

William A. Dossett. Born 1966. Admitted to bar: Minnesota, 1993. *Education:* Carleton College (B.A., 1989); University of North Carolina at Chapel Hill (M.R.P. 1993; J.D., 1993). (612-340-8765)

Gregory A. Fontaine. Born 1956. Admitted to bar: Minnesota 1981. *Education:* Northwestern University (B.A., 1978); University of Minnesota (J.D., 1981). Mr. Fontaine clerked for the Honorable Gerald W. Heaney, U.S. Court of Appeals, 8th Circuit, 1981-1982. *Publications*: Practical Pointers for Refinancing Property in a Contaminated Environment, Bureau of National Affairs (Jan. 1995); The Resource Conservation and Recovery Act in 1985, co-author, 3 Wm. Mitchell Envtl. L.J. 87 (1985). (612-340-8729)

Elizabeth A. Goodman. Born 1950. Admitted to bar: Michigan, 1977; Minnesota, 1978. *Education:* Alma College (B.A., 1972); University of Michigan (J.D., 1977). (Real Property Specialist Certified by the Real Property Section, Minnesota State Bar Association). *Publications*: New Guidelines May Streamline Environmental Assessments, Minnesota Ventures, 62 (Sept.-Oct. 1993); CERCLA's (Insecure) Secured Creditor Exemption, The Practical Real Estate Lawyer, 85 (May 1991). (612-340-2977)

Mark R. Kaster. Born 1957. Admitted to bar: Minnesota, 1984. *Education:* Unviersity of Minnesota (B.S. 1979; M.S., 1982); Marshall-Wythe School of Law, Exeter, England; William Mitchell College of Law (J.D., 1984). *Publications*: Minnesota Air Quality Handbook, co-editor (1992). (612-340-7815)

Alexandra B. Klass. Born 1967. Admitted to bar: Wisconsin, 1992; Minnesota, 1993. *Education:* University of Michigan (B.A., 1988); University of Wisconsin (J.D., 1992). Ms. Klass clerked for the Honorable Barbara B. Crabb, Chief Judge, U.S. District Court, Western District of Wisconsin, 1992-1993. (612-343-2166)

Thomas S. Llewellyn. Born 1955. Admitted to bar: District of Columbia, 1980. *Education:* University of Virginia (B.A., 1977); George Washington University (J.D., 1980). Mr. Llewellyn was Trial Defense Counsel, U.S. Army Judge Advocate General's Corp., 1981-1984. (202-452-6900)

Kevin J. Saville. Born 1964, Mason City, Iowa. Admitted to bar: Minnesota, 1992. *Education:* University of Northern Iowa (B.A., 1987); University of Minnesota (J.D., 1992). *Publications*: Discharging CERCLA Liability in Bankruptcy: When Does a Claim Arise?, 76 Minn. L. Rev. 327 (1991). (612-343-8215)

George Y. Sugiyama. Born. 1949. Admitted to bar: Maryland, 1976; District of Columbia, 1989. *Education:* Drake University (B.A., 1970); University of Maryland (J.D., 1976). Mr. Sugiyama served as an attorney at the U.S. Environmental Protection Agency, Washington, D.C. on Clean Air Act issues during 1976-1986. (202-452-6900)

CHAPTER ONE

AGENCIES WITH RESPONSIBILITY FOR ENVIRONMENTAL PROGRAMS

1.0 Environmental Regulation in Iowa

Iowa has an extensive system of environmental statutes and regulations. In particular, Iowa statutes have created programs that regulate discharges of pollutants into air, water, and groundwater and that regulate numerous specific activities concerning hazardous substances, solid waste, pesticides, fertilizers, and other chemicals. Iowa environmental laws impose additional obligations on landowners.

Iowa's voters have communicated to their public officials a strong interest in aggressive formulation and enforcement of environmental laws. Given the trend on the federal level toward criminal enforcement of environmental laws, persons subject to Iowa environmental laws and regulations should be sensitive to the public conviction behind those laws and the resulting influence on the attitudes of Iowa's lawmakers and environmental regulators at the state and local levels. This handbook discusses the statutes and regulations that make up Iowa's environmental law beginning with an introduction to the governmental bodies that administer this law.

1.1 United States Environmental Protection Agency

At the federal level, the United States Environmental Protection Agency (EPA) is the primary governmental agency responsible for implementing and enforcing environmental laws. The EPA headquarters is in Washington, D.C., but many EPA programs are operated from the EPA regional offices. Iowa is in EPA Region VII, which is administered from Kansas City:

U.S. Environmental Protection Agency Region VII
726 Minnesota Avenue
Kansas City, KS 66101
913-551-7000 -- Spill Reporting: 913-236-3778

EPA administers the hazardous waste management program in Iowa. As a result, EPA maintains inspection offices in Des Moines and Iowa City. In addition, EPA has maintained staff in Des Moines with responsibility for certain pesticide management program requirements.

The Pollution Prosecution Act of 1990, Pub. L. 101-593, 104 Stat. 2962, added fifty civil investigators to the EPA's Office of Enforcement staff to develop and prosecute civil and administrative actions. Moreover, the Pollution Prosecution Act increased the number of criminal investigators assigned to the EPA Office of Enforcement as follows:

1992	1993	1994	1995	1996
72	110	123	160	200

As you would expect, the number of criminal enforcement actions have grown proportionately. In 1993, EPA recovered a record $25.5 million in criminal fines and 943 months of jail time were imposed for criminal violations of environmental, health and safety (EHS) laws. The EPA relies on attorneys from the U.S. Department of Justice for a major part of EPA's legal work. As shown below, EPA's referrals to the Department of Justice for criminal prosecutions and EPA's criminal and administrative penalty enforcement actions taken have grown substantially over the past three years.

	1991	1992	1993
Criminal Prosecutions	81	107	140
Enforcement Actions	1755	1935	2010

On November 16, 1993, the Advisory Group on Environmental Offenses proposed that the U.S. Sentencing Commission adopt additional sentencing guidelines applicable to organizations for environmental criminal violations.[1/] These draft sentencing guidelines proposed that environmental crimes be punished more severely than most other crimes. While the proposed guidelines may be subject to certain revisions before taking effect, the sentencing guidelines ultimately adopted will likely be based on the proposed guidelines and should be evaluated in connection with environmental enforcement efforts.

1.2 Department of Natural Resources

The Iowa Department of Natural Resources (DNR) is the state agency with primary responsibility for protecting the environment and managing energy, fish, wildlife, land, and water resources in the state.[2/] The DNR director is appointed by the Governor, subject to confirmation of the senate. The DNR is headquartered at:

Iowa Department of Natural Resources
Wallace State Office Building
900 East Grand Avenue
Des Moines, Iowa 50319-0034
515-281-5145

The DNR also maintains regional offices in Manchester, Mason City, Spencer, Atlantic, and Washington, Iowa. These offices are responsible for local investigations and activities.

The DNR establishes policies and adopts rules pursuant to the Iowa Administrative Procedure Act (APA),[3/] primarily through the Natural Resource

---

1/ 58 Fed. Reg. 65,764 (Dec. 16, 1993).

2/ Iowa Code § 455A.2.

3/ Iowa Code ch. 17A.

Commission (NRC) and the Environmental Protection Commission (EPC). These Commissions are attached to the DNR, make policy for major segments of the DNR's programs, and may adjust the final DNR budget request submitted to the Governor's office:

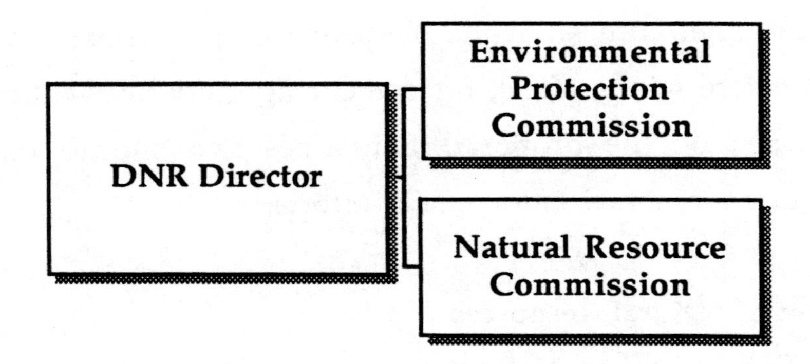

1.2.1 Natural Resource Commission

The NRC consists of seven members appointed by the Governor and deals mainly with hunting, fishing, and outdoor recreation issues.[4/] The NRC has authority to establish policy and adopt rules relating to protected water areas, endangered species, public lands and waters, wildlife conservation, and other designated subjects. The NRC has responsibility for hearing appeals in contested cases realting to these issues. The NRC must approve or disapprove proposals of the DNR director for the acquisition or disposition of state lands and waters relating to state parks, recreational facilities, and wildlife programs.

1.2.2 Environmental Protection Commission

The EPC consists of nine members appointed by the Governor and handles issues related to water, air, and waste management.[5/] The EPC adopts rules and

---

4/ Iowa Code § 455A.5; see also Marksbury v. State, 322 N.W.2d 281 (Iowa 1982) (regarding jurisdictional authority over beach on Lake Okoboji).

5/ Iowa Code § 455A.6; see also Cota v. Iowa Envtl. Prot. Comm'n, 490 N.W.2d 549 (Iowa 1992).

establishes policy regarding environmental protection issues under the Iowa Environmental Quality Act, Iowa Code ch. 455B, the bottle bill, waste reduction and recycling, groundwater protection, household hazardous waste, and milldams and raceways. The EPC also hears contested cases in matters relating to these issues in accord with the Iowa APA. As with NRC contested cases, these contested cases may be delegated to be heard by an administrative law judge (ALJ) from the Department of Inspections and Appeals. The EPC also approves the budget request for the DNR on environmental protection issues.

1.2.3 DNR Enforcement and Penalties

Implementation of DNR environmental protection programs is carried out by DNR staff, primarily in the Environmental Protection Division of the DNR:

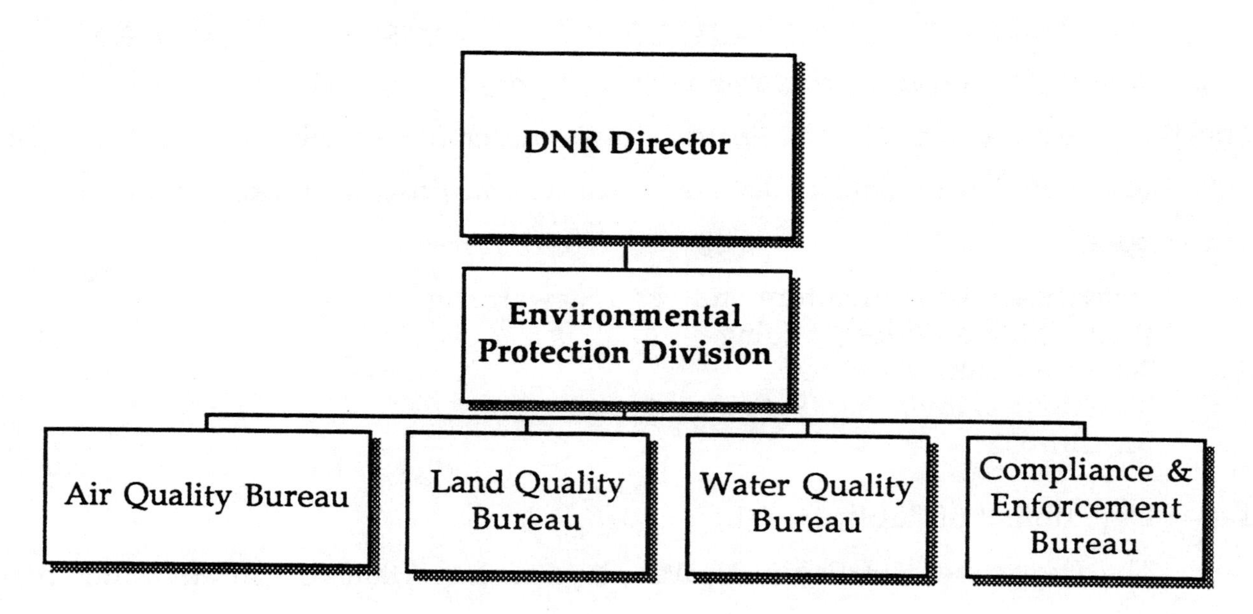

In addition to referring cases for enforcement action to the Attorney General as discussed below, the DNR may issue administrative penalties for violations of the Iowa Environmental Quality Act, Iowa Code chapter 455B. No administrative penalty may be assessed unless the cumulative amount is $50 or more.[6/] Among other factors, the DNR applies the following criteria for screening and assessing administrative penalties:

1. Costs saved by noncompliance;
2. Gravity of the violation;
3. Culpability;
4. Statutory maximum penalties.[7/]

1.3 Department of Agriculture and Land Stewardship

The Department of Agriculture and Land Stewardship (DALS) also has responsibility for several programs relating to environmental protection in Iowa. The DALS administers Iowa's pesticide laws under Iowa Code chapter 206. The DALS also handles the soil and water conservation program. The DALS may be contacted at:

Department of Agriculture and Land Stewardship
Wallace State Office Building
900 East Grand Avenue
Des Moines, Iowa 50319
515-281-5321

1.4 Department of Public Health

The Department of Public Health plays an important role in environmental health issues in Iowa. For example, the Department of Public Health is responsible

---

6/ Iowa Admin. Code § 567-10.3(1).

7/ Iowa Code § 455B.109; Iowa Admin. Code § 567-10.2. Later chapters in this handbook discuss specific enforcement issues applicable to the subjects in those chapters.

for the lead abatement program,[8/] aspects of the asbestos health protection requirements, and certain water quality standards. The Department of Public Health also administers the PCB program in cooperation with EPA. Finally, the Department of Public Health cooperates with the Commissioner of Labor on certain occupational safety and health act (OSHA) issues. The Department of Public Health may be contacted at:

Department of Public Health
Lucas State Office Building
321 East 12th Street
Des Moines, Iowa 50319-0075
515-281-5787

1.5 Department of Employment Services

The Department of Employment Services, headed by the Commissioner of Labor, maintains a Division of Labor Services that is responsible for implementing a major portion of the hazardous chemicals right-to-know law.[9/] The Iowa Commissioner of Labor has authority to adopt and enforce rules regarding communication of hazards in the workplace, including information about hazardous chemicals to which an employee may be exposed in the workplace. In cooperation with the Department of Public Health and the DNR, the Labor Commissioner also may expand the list of hazardous chemicals or reporting required under the right-to-know law. The Labor Commissioner also certifies asbestos removal contractors in Iowa. The Department of Employment Services may be contacted at:

Department of Employment Services
1000 East Grand Avenue
Des Moines, Iowa 50319
515-281-3606

8/ Iowa Code §§ 135.101-135.105; see also chapter 13 of this handbook regarding asbestos, PCB, and lead regulation.

9/ Iowa Code ch. 89B; see also chapter 8 of this handbook regarding right-to-know laws.

1.6 Department of Public Safety

The Department of Public Safety's Division of State Fire Marshal maintains responsibility for aboveground storage tank regulation.[10/] While the DNR administers the underground storage tank (UST) program in Iowa, the Fire Marshal also may become involved in UST issues. The Fire Marshal may be contacted at:

Division of State Fire Marshal
Department of Public Safety
Wallace State Office Building
900 East Grand Avenue
Des Moines, Iowa 50319-0047
515-281-5821

1.7 Attorney General's Office

The Iowa Attorney General's office maintains an Environmental Law Division responsible for enforcing Iowa's environmental laws on behalf of various state agencies. At the request of the DNR Director, for example, the Attorney General's office institutes legal proceedings necessary to enforce hazardous waste management and other permit requirements.[11/] Similar enforcement authority applies to most environmental laws administered by the various state agencies.

In addition to the duty to commence legal proceedings at the request of the DNR, EPC or other state agencies, the attorney general may institute civil or criminal proceedings, including an action for injunction, to enforce the provisions of and orders or rules issued under authority of the Iowa Environmental Quality Act, Iowa Code ch. 455B, which includes most of the basic environmental laws.[12/]

10/ Iowa Code ch. 101; see also chapter 10 of this handbook regarding storage tank regulation.

11/ Iowa Code § 455B.418(d).

12/ Iowa Code § 455B.112.

The Iowa Attorney General's office may be reached at:

Office of the Attorney General
Environmental Law Division
Executive Hills East, 2d floor
1223 East Court Avenue
Des Moines, Iowa 50319
515-281-5351

1.8 County Resource Enhancement Committees

In 1989, the Iowa legislature adopted an act stating the long-term environmental policy of the state:

> It is the policy of the state of Iowa to protect its natural resource heritage of air, soils, waters, and wildlife for the benefit of present and future citizens with the establishment of a resource enhancement program. The program shall be a long-term integrated effort to wisely use and protect Iowa's natural resources through the acquisition and management of public lands; the upgrading of public park and preserve facilities; environmental education, monitoring, and research; and other environmentally sound means.13/

To carry out this program, each county in Iowa has a County Resource Enhancement Committee. These committees are intended to coordinate resource enhancement programs, plans, and proposed projects developed by cities, county conservation boards, and soil and water conservation district commissioners.14/

1.9 Comprehensive Petroleum Underground Storage Tank Fund Board

The Comprehensive Petroleum Underground Storage Tank Fund Board (Petrofund Board) manages a program for the cleanup of petroleum that leaks from underground storage tanks.15/ This program involves the administration of a statutorily created fund based on a tax imposed on petroleum products stored in

---

13/ Iowa Code § 455A.16.

14/ Iowa Code § 455A.20.

15/ See Iowa Code §§ 455G.1-455G.20; see also chapter 10 of this handbook regarding storage tank regulation.

tanks. Persons who are responsible for cleaning up a leak of petroleum products from a petroleum tank may apply to the Board for reimbursement of certain expenses. This program and the related insurance program are currently administered by:

UST Financial Responsibility Program
Williams & Co.
1000 Illinois Street, Suite B
Des Moines, IA 50314
515-284-1616

1.10 Center for Health Effects of Environmental Contamination

The University of Iowa maintains the Center for Health Effects of Environmental Contamination. The Center's purpose is to determine the levels of environmental contamination which can be specifically associated with human health effects.[16/] The Center provides a valuable source of data related to contamination in soil, air, water, and food.

1.11 Center for Agricultural Health and Safety

The Center for Agricultural Health and Safety is a joint venture of the University of Iowa and Iowa State University formed by the legislature to establish farm health and safety programs. These programs are intended to reduce the incidence of disabilities suffered by persons engaged in agriculture which results from disease or injury.[17/] The Center originated as a pilot program at the University of Iowa College of Medicine to provide comprehensive health and safety services and cooperates with the Center for Health Effects of Environmental Contamination on certain projects.

16/ Iowa Code § 263.17.

17/ Iowa Code § 262.78.

1.12 County and City Agencies

County and city ordinances frequently impose special environmental obligations on individuals and corporations, including license and permit requirements. These ordinances are interpreted and enforced by various local governmental bodies, including county boards of commissioners, city councils, and local environmental agencies.[18]

For example, County Attorney Offices have authority, in cooperation with the Iowa Attorney General's Office, to commence criminal prosecution of hazardous waste violations pursuant to Iowa Code § 716B.5. City Fire Marshals and local fire departments have responsibilities related to toxic chemicals under the Community Right-to-Know laws discussed in chapter 8 of this handbook. Applicants for a sanitary landfill or an infectious waste incinerator must obtain siting approval from local authorities.[19] When faced with an environmental question, be sure to consider the obligations that may be imposed by these governmental bodies.

18/ See State v. Butler, 419 N.W.2d 362 (Iowa 1988) (upholding local health board's authority over sewage disposal pursuant to Iowa Code § 137.21).

19/ Iowa Code § 455B.305A; see also chapter 7 of this handbook regarding infectious waste management and chapter 9 of this handbook regarding solid waste management.

CHAPTER TWO

AIR POLLUTION CONTROL

2.0 Introduction

Iowa has an extensive system of air quality laws and regulations to protect public health, welfare, safety, and the natural resources of the state. The Department of Natural Resources (DNR) is the state agency with primary authority to prevent, abate, or control air pollution in Iowa.[1]

As in most states, the air pollution regulatory scheme in Iowa is intended to carry out and to supplement the federal air pollution control program under the federal Clean Air Act, which was enacted in 1970 and was substantially expanded in 1990.[2] The U.S. Environmental Protection Agency (EPA) has delegated to the state primary responsibility for assuring air quality. The EPA, however, maintains jurisdiction and oversight authority to enforce the Clean Air Act. This chapter describes Iowa's laws and regulations governing air pollution control.

2.1 Iowa Code Chapter 455B -- Authority for Air Pollution Control

The Iowa legislature has adopted Iowa Code chapter 455B, Division II, which established broad authority for the control of air contamination and pollution.[3] The legislature authorized the Environmental Protection Commission (EPC), acting under the umbrella of the DNR, to "[d]evelop comprehensive plans for the abatement, control, and prevention of air pollution in this state, recognizing varying requirements for different areas in the state."[4] Moreover, the legislature authorized the EPC to adopt the following:

1. Rules necessary to obtain approval of the state implementation plan (SIP) under the federal Clean Air Act;

1/ Iowa Code § 455B.132.

2/ 42 U.S.C. §§ 7401-7671q.

3/ Iowa Code §§ 455B.131-455B.151.

4/ Iowa Code § 455B.133(1).

2. Ambient air quality standards to protect the public health and welfare and to reduce emissions contributing to acid rain pursuant to Title IV of the federal Clean Air Act;

3. Emission limitations or standards relating to the maximum quantities of air contaminants that may be emitted from any air contaminant source;

4. Classifications of air contaminant sources;

5. Rules requiring notice of the construction of any air contaminant source; and

6. Rules requiring the owner or operator of an air contaminant source to obtain an operating permit prior to operation of the source (Title V operating permit).[5]

The air quality rules adopted by the EPC are codified at Iowa Admin. Code chapters 567-20 to 567-30.

2.2 DNR Air Quality Bureau

The DNR's Air Quality Bureau administers and enforces all federal and state laws relating to air pollution in Iowa. These laws aim at meeting air quality standards in nonattainment areas[6] and prevention of significant deterioration (PSD) in those areas which meet national ambient air quality standards or are unclassified.[7] The DNR monitors air quality at sites throughout the state and enforces pollution control standards primarily by means of environmental permits.[8] The permits issued by the agency contain specific enforceable conditions and limitations for compliance with federal and state air quality standards. The

---

5/ Iowa Code § 455B.133.

6/ Iowa Admin. Code § 567-22.5.

7/ Iowa Admin. Code § 567-22.4.

8/ Iowa's air quality control regions are designated at 40 C.F.R. § 81.316.

DNR implements the requirements of the federal Clean Air Act in Iowa through a state implementation plan or SIP and written cooperative agreements with the EPA.9/ Finally, the agency routinely inspects permitted facilities to assure compliance with air quality laws and regulations. If violations are discovered, the DNR may refer the matter to the state attorney general's office for enforcement.

The EPC has recently adopted new permit rules to meet the Title V provisions of the federal Clean Air Act in compliance with 40 C.F.R. part 70.10/ Under these rules, all Iowa facilities that require an operating permit must submit permit applications to the DNR.

2.3 Iowa Standards for Air Pollution Control

The DNR groups its air quality standards and methods of measurement at Iowa Admin. Code chapters 567-23, 567-25, and 567-28. The standards contained in these parts must be met as part of a facility operating permit. No person may emit any pollutant in such an amount or in such a manner as to cause or contribute to a violation of any ambient air quality standard beyond the person's property line, provided that in the event the general public has access to the person's property or portion thereof, the ambient air standards shall apply in all locations.11/ Further, no person may permit the handling, use, or storage of any material in a manner which allows avoidable amounts of particulate matter to become airborne.12/ Excess emissions must be reported.13/ More stringent rules apply during air pollution emergency episodes.14/

9/ Iowa's SIP is set forth in 40 C.F.R. §§ 52.820-52.833.

10/ 42 U.S.C. §§ 7661-7661f; see Iowa Admin. Code §§ 567-22.100 to 567-22.116.

11/ Iowa Admin. Code §§ 567-21.1, 567-23.1.

12/ Iowa Admin. Code §§ 567-21.1, 567-23.3.

13/ Iowa Admin. Code § 567-24.1.

14/ Iowa Admin. Code §§ 567-26.1 to 567-26.4.

Air Pollution Control

The Iowa standards for air pollution control are premised on achieving attainment of the following:

1. National Ambient Air Quality Standards (Sulfur dioxide (SO_2), Nitrogen oxides (NO_x), Carbon monoxide (CO), Particulates (PM_{10}), Ozone, and Lead);
2. National Emission Standards for Hazardous Air Pollutants (NESHAPs);
3. Title V or Part 70 permits for major stationary sources;
4. New Source Performance Standards (NSPS);
5. Acid deposition control; and
6. Mobile source control.

2.3.1 Attainment of National Ambient Air Quality Standards (NAAQS)

The federal Clean Air Act establishes the National Ambient Air Quality Standards (NAAQS).[15] The goal of the Act is to protect the health and welfare of the public by achieving and maintaining these standards. Each air quality standard is the sum of experimentally-determined injury threshold concentrations, plus an adequate margin of safety. Sensitivity thresholds to air pollution vary with the population segment being considered. For example, individuals with pulmonary and respiratory diseases, children, and the elderly are particularly sensitive to pollutant exposures. These population segments are of primary concern in the setting of air quality standards.

There are two levels of air quality standards established: primary and secondary.[16] The primary standards are set at concentrations which are ascertained to be low enough to protect the health of the public. Physiological changes due to air pollutants that interfere with normal activity in sensitive or healthy individuals are the most important considerations in establishing primary standards. Secondary

---

15/ 42 U.S.C. § 7409.

16/ 42 U.S.C. § 7409.

standards are established to protect public welfare and take into account concentration levels that would cause injury to livestock and agricultural crops, damage to property, annoyance, and transportation hazards.

While both primary and secondary air quality standards are established at the federal level, the Clean Air Act grants states the authority to set standards which are more stringent than the federal standards. Determination of maximum concentration levels for setting state standards is the responsibility of the EPC and the DNR.

Iowa's ambient air quality standards are set forth in Iowa Admin. Code chapter 567-28. The status of Iowa programs for attainment of NAAQS is set forth in 40 C.F.R. § 81.316. Iowa has adopted special requirements for major stationary sources in areas designated as attainment areas that are intended to prevent significant deterioration of air quality and to keep those areas in compliance with the NAAQS.[17] More stringent requirements apply in nonattainment areas in an effort to bring those areas of the state in compliance with the NAAQS.[18]

2.3.2 Iowa Pollution Control Performance Standards Governing Specific Activities and Pollutants

In addition to attainment of the NAAQSs, Iowa regulations concentrate on pollution control for air emissions from industrial facilities. Iowa has adopted a plan for control of designated pollutants from certain existing facilities pursuant to the Clean Air Act § 111(d). This plan applies to certain existing sulfuric acid production plants and phosphate fertilizer plants.[19]

17/ Iowa Admin. Code § 567-22.4.

18/ Iowa Admin. Code § 567-22.5.

19/ 40 C.F.R. §§ 62.3850-62.3911.

For other specific processes, the EPC has established for the DNR performance standards to control criteria pollutants.[20] Special provisions apply to anaerobic lagoons.[21]

2.3.3 Control of Incinerators and Open Burning

The EPC has adopted rules for the DNR regulating the opacity and particulate content of emissions from sources in the state.[22] These rules are often significant in the control of incinerators and similar sources. In 1993, the Iowa legislature enacted a moratorium on the construction of commercial waste incinerators. This legislation prohibits the DNR from granting a permit for the construction or operation of a commercial waste incinerator until such time as the DNR or the EPA adopts rules governing "safe emission standards for releases of toxic air emissions from commercial waste incinerators."[23] This moratorium primarily applies to certain toxic or hazardous waste incinerators, but does not apply to waste oil burners.[24]

Iowa generally prohibits open burning.[25] This general prohibition is subject to several exceptions, and there is an application procedure for persons who want a permit for open burning. Nonetheless, municipalities may impose a total prohibition on open burning.

20/ Iowa Admin. Code § 567-23.4.

21/ Iowa Admin. Code § 567-23.5.

22/ Iowa Admin. Code § 567-29.1; see also 42 U.S.C. § 7429 (regarding new source performance standards for solid waste incineration units).

23/ Iowa Code § 455B.151; see also 59 Fed. Reg. 48,198 (Sept. 20, 1994) (proposing performance standards for municipal waste combustors). Chapter 9 of this handbook discusses solid waste management regulation.

24/ Iowa Code § 455B.151(2)(c).

25/ Iowa Admin. Code § 567-23.2.

2.3.4 Iowa Acid Rain Program

In the 1990 amendments to the federal Clean Air Act, Congress adopted extensive provisions designed to reduce emissions contributing to acid rain, which are generally referred to as Title IV of the federal Clean Air Act.26/ These provisions and EPA's implementing regulations primarily aim to reduce emissions of sulfur dioxide and nitrogen oxides.27/

Iowa has adopted rules restricting sulfur dioxide and nitrogen oxide emissions based on the Title IV requirements.28/ The acid rain program includes permitting requirements similar to the Title V operating permit program discussed in more detail below. Like the Title V operating permits, acid rain control permits may include a permit shield, which provides that compliance with the permit will be deemed compliance with any applicable air pollution control requirements.29/ In addition, continuous emission monitoring (CEM) requirements apply to certain potential acid rain sources.30/

2.3.5 Iowa Air Toxics Program

Beginning July 1, 1991, Iowa imposed an annual temporary air toxics fee of $25 per ton of hazardous air pollutants (HAP) emitted.31/ This program was intended to raise money for the DNR to prepare, submit, and obtain approval of the Title V operating permit program.32/ With the adoption of Iowa's Title V operating

---

26/ 42 U.S.C. §§ 7651-7651o.

27/ 40 C.F.R. pts. 760, 762, 763; see also 59 Fed. Reg. 13538 (Mar. 22, 1994).

28/ Iowa Admin. Code §§ 567-22.120 to 567-22.147.

29/ Iowa Admin. Code § 567-22.134; see also section 2.4.9 of this handbook.

30/ Iowa Admin. Code § 567-25.2.

31/ Iowa Code § 455B.133A; Iowa Admin. Code ch. 567-30.

32/ Iowa Code § 455B.133B.

permit regulations, the temporary air toxics fees is being phased out.33/ Iowa has promulgated emission standards applicable to certain hazardous air pollutants.34/

The federal program is expected to establish maximum achievable control technology (MACT) to be applied to reduce emissions of 186 statutorily listed hazardous air pollutants and other hazardous air pollutants that EPA may add to the list.35/ The federal standards are being phased in, and the DNR is expected to adopt similar state standards in turn.

2.4 Iowa Air Quality Permit Program

There are three primary air emission permit programs under the new Iowa air quality rules. These include the following:

1. Construction permits;
2. Title V operating permits; and
3. Acid rain deposition permits.

The construction permit program has been in place for a number of years. The operating permit and acid rain permit programs were adopted in 1994 following the 1990 Clean Air Act amendments and continue to evolve. Because the Title V operating permit program is the most extensive, this chapter will focus on the Title V operating permit program after a brief discussion of the construction permit program.

2.4.1 Construction Permits and Emissions Trading

The DNR requires the owner or operator of a facility to obtain a construction permit prior to construction or modification of a facility that may contribute to air

---

33/ Iowa Admin. Code §§ 567-22.100 to 567-22.116.

34/ Iowa Admin. Code § 567-23.1.

35/ 42 U.S.C. § 7412. With regard to risk management plans to prevent accidental releases of extremely hazardous chemicals required by 42 U.S.C. § 7412(r), see section 8.1.4 of this handbook.

pollution.[36] In attainment or unclassified areas (areas presumed to have attained NAAQS), the DNR will grant a construction permit if operation of the facility will not cause significant deterioration in air quality. In non-attainment areas (areas with air pollution exceeding NAAQS), the facility owner or operator must offset the emissions from the new facility. Offsets may be obtained through the emissions trading program[37] or by reducing emissions from the another part of the same facility under the "bubble" program.[38] A facility can earn offsets or Emission Reduction Credits (ERC), for example, by changing to a cleaner fuel source, reducing operating hours, or other enforceable pollution control measures. Emissions trading also may be applied to permits other than construction permits.

The DNR publishes notice of proposed construction permits and accepts public comments on proposed construction permits before making a permit decision.[39] The DNR may impose conditions on construction permits.[40]

2.4.2 Title V Operating Permits

The general rules for obtaining a Title V operating permit are found in Iowa Admin. Code §§ 567-22.100 to 567-22.116. Under these rules, no source may operate after the time that it is required to submit a timely and complete application for a Title V permit. The requirement to obtain a Title V operating permit is deferred until 1999 except for major sources, affected sources, and solid waste incineration units governed by the Clean Air Act § 129(e).[41] Except for certain designated "insignificant" sources, Title V operating permits are required for the following:

1. Any affected source subject to Title IV acid rain provision;

36/ Iowa Code 455B.133(6); Iowa Admin. Code §§ 567-22.1(1), 567-22.4.

37/ Iowa Admin. Code §§ 567-22.5(2)-(4), 567-22.7.

38/ Iowa Admin. Code § 567-23.6.

39/ Iowa Admin. Code § 567-22.2(2).

40/ Iowa Admin. Code § 567-22.3.

41/ Iowa Admin. Code § 567-22.101(2).

2. Any major source;

3. Any source subject to New Source Performance Standards in Iowa Admin. Code § 567-23.1(2) or NESHAPs regulations pursuant to the Clean Air Act § 112 or Iowa Admin. Code § 567-23.1(3);

4. Solid waste incineration units governed by the Clean Air Act § 129(e);

5. Any source category designated by the EPA under 40 C.F.R. § 70.3.[42/]

This list is a substantial expansion over the number of facilities required to obtain an air permit prior to the 1990 Clean Air Act amendments.

The DNR permit application process requires preparation of a very detailed application covering items set forth in Iowa Admin. Code § 567-22.105. In addition to initial applications, a supplemental application must be submitted six months prior to any planned significant modification of a Title V permit.[43/]

Any application form, report, or compliance certification must contain certification by a responsible corporate official of the truth, accuracy, and completeness of its contents based on information and belief formed after a reasonable inquiry.[44/] Annual certificates of compliance must be submitted by a responsible corporate official.[45/] Enforcement agencies may use these certifications in an effort to hold corporate officers personally responsible for compliance issues.

2.4.3 Title V Operating Permit Conditions And Monitoring Requirements

The DNR must issue permits which include emission limitations, operational requirements, and other provisions needed to ensure compliance with all applicable requirements at the time of permit issuance.[46/] The DNR imposes a

---

42/ Iowa Admin. Code § 567-22.102. These terms are defined at length in Iowa Admin. Code § 567-22.100.

43/ Iowa Admin. Code §§ 567-22.105(1)(a)(4), 567-22.113.

44/ Iowa Admin. Code § 567-22.107(4).

45/ Iowa Admin. Code § 567-22.105(2)(i).

46/ Iowa Admin. Code § 567-22.108(1).

permit fee of $24 per ton of actual emissions of each regulated air pollutant.[47/] Even when an operating permit has been issued, the construction permit requirement remains applicable.[48/]

All permits must require the permittee to comply with emissions monitoring and analysis procedures or test methods required under the applicable requirements,[49/] including any procedures and methods promulgated pursuant to the federal Clean Air Act.[50/]

2.4.4 Recordkeeping Requirements

The agency, at a minimum, must include the following recordkeeping requirements in all permits:[51/]

1. Maintain records adequate to document compliance at the stationary source, including detailed sampling and analytical information; and

2. Retain records at the stationary source of all monitoring data and support information for a period of five years, or longer as specified by the commissioner, from the date of the monitoring sample, measurement, or report.

2.4.5 Reporting Requirements

All required monitoring reports must be submitted at least every six months.[52/] In the event of any deviation from permit requirements, the permittee must promptly notify the DNR of the deviation, its probable cause, and steps taken or planned to correct and prevent reoccurrence of the deviation.[53/] As noted above,

---

47/ Iowa Admin. Code § 567-22.106.

48/ Iowa Admin. Code §§ 567-22.1, 567-22.107(1)(g).

49/ Iowa Admin. Code § 567-22.108(3).

50/ See 42 U.S.C. §§ 7414 or 7661c(b).

51/ Iowa Admin. Code § 567-22.108(4).

52/ Iowa Admin. Code § 567-22.108(5)(a).

53/ Iowa Admin. Code § 567-22.108(5)(b).

annual compliance certificates must be submitted by a responsible corporate official.[54]

2.4.6 Alternative Operating Scenarios

The permit applicant may request the agency to permit reasonably anticipated alternative operating scenarios. The permit for alternative operating scenarios must (1) require the stationary source, contemporaneously with making a change from one operating scenario to another, to record in a log at the permitted facility a record of the scenario under which it is operating; and (2) ensure that the operation under each alternative operating scenario complies with all applicable air pollution control requirements and the requirements under the permit rule.[55]

2.4.7 Additional Title V Operating Permit Compliance Requirements

All DNR air permits contain inspection and entry requirements, a compliance schedule if required under the permit rule, and provisions establishing the permit shield discussed in section 2.4.9 of this handbook. In addition to the permit conditions outlined above, all DNR air emission permits must incorporate a number of general permit conditions set forth in Iowa Admin. Code § 567-22.108(15), (16).

2.4.8 Duration of Title V Operating Permits

The duration of air emission permits issued by the DNR is dependent on the specific permit. Most Title V operating permits are issued for periods of five years. The agency can void an existing permit if it determines that the stationary source no longer requires a permit. The agency has the authority to reopen a permit when additional federal air pollution control requirements become applicable to a

---

54/ Iowa Admin. Code § 567-22.105(2)(i).

55/ Iowa Admin. Code § 567-22.108(12).

stationary source with a remaining permit term of three or more years.56/ The permit can also be reopened on grounds of material mistake or endangerment to public health or the environment. The agency may also revoke a permit for a variety of circumstances including unresolved noncompliance, endangerment of health or the environment, and the permittee's failure to disclose relevant facts in the permit process.

2.4.9 Permit Shield

The DNR may include a provision in new permits called the permit shield, which provides that compliance with the conditions of the permit shall be deemed compliance with any applicable air pollution control requirements.57/ This provision is designed to insulate the permittee from air pollution violations as long as the permittee remains in compliance with its permit.

2.5 General Permits, Variances, and Permits By Default

In 1993, the Iowa legislature adopted amendments authorizing the DNR to issue general permits for classes of air contaminant sources that can be described and conditioned by a single general permit similar to the general permit for stormwater discharges discussed in chapter 3 of this handbook.58/ After the effective date of a general permit, a covered person must provide notice of intent to comply with the general permit. The DNR is expected to adopt general permits for some of the more common activities governed by the Title V operating permit program.

If the DNR fails to take action within 60 days after an application for a permit variance, the inaction will be deemed to be a grant of the variance.59/ Similarly, if

---

56/ Iowa Admin. Code § 567-22.108(17).

57/ Iowa Admin. Code § 567-22.108(18); see also Iowa Admin. Code § 567-22.134 (regarding acid rain permit shield).

58/ Iowa Code § 455B.103A.

59/ Iowa Code § 455B.147; Iowa Admin. Code § 567-21.2.

the DNR fails to enter a final order within 60 days after the final argument in an appeal hearing, the inaction will be deemed a finding favorable to the respondent in the hearing. This variance procedure, however, is inapplicable to applications for a conditional permit for an electrical power generating facility and has not been recognized yet in reported appellate case law.

For permit applications other than air permits, the DNR must either approve or deny the permit within six months from the date that the DNR receives a completed application. Inaction results in issuance of the permit by default.[60] The legislature excluded air permits from this permit by default law, and in 1993 the legislature removed a permit by default provision previously applicable to construction permits.

2.6 Federal Compliance and Enforcement

Enforcement sanctions available under federal and Iowa air pollution control legislation have been strengthened substantially by the 1990 Clean Air Act amendments and revisions to the Iowa Code. The heavy penalties that can be imposed, particularly as federal and state requirements become significantly more restrictive, make it essential that companies achieve timely compliance with new requirements and develop programs to maintain compliance.

The EPA and DNR have civil, criminal, and administrative enforcement tools under their respective enabling legislation. Both federal and state laws also allow private citizens to enforce or require enforcement of their respective air pollution control programs.

2.6.1 EPA Notices of Violation

When enforcement involves an alleged violation of a state implementation plan (SIP) or a permit, EPA must first issue a notice of violation (NOV) to the

---

60/ Iowa Code § 455B.104.

violating facility.[61/] The NOV typically identifies the violation and requires that it be corrected within a certain timeframe. Thirty days after the NOV is served, EPA may either issue an administrative order, file a civil action, or initiate an administrative penalty proceeding to compel compliance. An NOV is not a final order subject to judicial review, and EPA has discretion to determine what, if any, further enforcement action to take following issuance of the NOV.

2.6.2 EPA Administrative Orders

EPA may issue administrative orders to any person that has violated any requirement or prohibition of an applicable SIP or permit after issuance of an NOV. EPA may also issue administrative orders, without first providing an NOV, against persons that violate provisions such as new source performance standards, permit requirements, or acid rain requirements.[62/] These orders generally require the violator to undertake specific corrective actions within an established deadline. EPA also is authorized to issue orders prohibiting construction or modification of any major stationary source if the state in which the source is located has failed to comply with certain requirements of the Clean Air Act.[63/] EPA administrative orders must provide an opportunity for the source to confer with the agency about the alleged violation.[64/] Orders also must require compliance within one year after issuance, and they are not renewable. EPA may enforce its administrative orders through civil lawsuits seeking injunctive relief, monetary penalties, or both.[65/] Issuance of an order does not preclude EPA from simultaneously pursuing other available enforcement remedies.

61/ CAA § 113(a)(1), 42 U.S.C. § 7413(a)(1).

62/ CAA § 113(a)(3), 42 U.S.C. § 7413(a)(3).

63/ CAA § 113(a)(5), 42 U.S.C. § 7413(a)(5).

64/ CAA § 113(a)(4), 42 U.S.C. § 7413(a)(4).

65/ CAA § 113(a), 42 U.S.C. § 7413(a).

2.6.3 EPA Civil Judicial Penalties

In addition to its administrative enforcement options, EPA may commence a civil lawsuit against: (1) any person who has violated applicable SIP or permit or other requirements under the Clean Air Act; (2) any person attempting to construct or modify a major stationary source in violation of prevention of significant deterioration (PSD) or nonattainment requirements; (3) or any person who has violated certain administrative orders.[66] EPA may seek a temporary or permanent injunction against the violators, civil penalties of up to $25,000 per violation per day, or both.

The Clean Air Act's civil penalty provision establishes a strict liability standard. If a facility is not in full compliance with applicable requirements, it is subject to these civil penalties regardless of the good faith efforts of the facility's operator to comply with applicable requirements.

2.6.4 EPA Administrative Penalties

EPA has authority to assess administrative penalties without the need for time-consuming court proceedings seeking the civil remedies outlined in the preceding section.[67] EPA may impose administrative penalties for the same violations to which civil penalties apply. The administrative penalties may be up to $25,000 per day per violation, subject to a maximum penalty of $200,000.[68] A person is entitled to an administrative hearing on the record under the federal Administrative Procedure Act before EPA may impose any administrative penalties.[69] If the penalized party fails to pay, EPA may bring a civil action to recover the amount of penalty, an additional "nonpayment penalty" of 10 percent

---

66/ CAA § 113(b), 42 U.S.C. § 7413(b).

67/ CAA § 113(d), 42 U.S.C. § 7413(d).

68/ CAA § 113(d)(1), 42 U.S.C. § 7413(d)(1).

69/ CAA § 113(d)(2)(A), 42 U.S.C. § 7413(d)(2)(A).

per quarter, plus interest and enforcement expenses.[70/] The penalized party can seek judicial review, but the legal standard to overturn the penalties, which requires a showing that there is no substantial evidence supporting the agency's action or there has been an abuse of discretion, is difficult to meet.[71/]

In addition, EPA is authorized to develop a program allowing issuance of so-called "field citations" assessing penalties of up to $5,000 per day for minor violations under certain circumstances.[72/] Recipients of such citations can either pay the administrative fine or request an informal hearing. EPA also is authorized to assess administrative "noncompliance" penalties in some circumstances that are intended to approximate the economic gain or benefit realized by the violator as a result of its noncompliance.[73/]

2.6.5 EPA Criminal Enforcement

EPA also may seek criminal penalties for "knowing" violations of virtually every requirement of the Clean Air Act. Under the 1990 Amendments, criminal violations have been upgraded from misdemeanors to felonies. The criminal sanctions apply to "any responsible corporate officer as well as the entity operating the facility."[74/] Courts may impose criminal responsibility only upon persons with "actual awareness or actual belief" that a violation is taking place. However, if a person takes steps to be shielded from relevant information, he or she may still be

---

70/ CAA § 113(d)(5), 42 U.S.C. § 7413(d)(5).

71/ CAA § 113(d)(4), 42 U.S.C. § 7413(d)(4); see also 5 U.S.C. § 706.

72/ CAA § 113(d)(3), 42 U.S.C. § 7413(d)(3).

73/ CAA § 120, 42 U.S.C. § 7420; section 2.6.7 of this handbook discusses how EPA determines economic benefit.

74/ CAA § 113(c)(6), 42 U.S.C. § 7413(c)(6).

held criminally liable.[75] Criminal sanctions also may be invoked against lower-level employees who knowingly cause a violation.[76]

The maximum penalties for each knowing violation of a SIP requirement, an administrative order, and certain other listed requirements include criminal fines of up to $250,000 ($500,000 for corporations), imprisonment for up to five years, or both, with each day of noncompliance constituting a separate violation.[77] In addition, a criminal fine of up to $250,000 ($500,000 for corporations) and/or imprisonment for up to two years may be imposed against any person knowingly making a false material statement, representation, certification, or omission in any document required to be filed or maintained; altering, concealing, or failing to maintain or file such a document; failing to make a required notification or report; or tampering with or failing to install a monitoring device.[78] Knowing failure to pay a fee required under the Clean Air Act may give rise to a fine of up to $250,000 ($500,000 for corporations), imprisonment for up to one year, or both. These criminal sanctions generally can be doubled for the second and subsequent violations.

There are additional criminal provisions for hazardous emissions that create "imminent danger of death or serious bodily injury." If the emissions release is negligent, the maximum penalty is a fine of up to $250,000 ($500,000 for corporations), imprisonment of up to one year for each violation, or both, with the penalty doubled after the first offense.[79] If the release is knowing, the maximum penalty is a fine of up to $250,000 ($1 million for corporations) and imprisonment of up to 15 years for each violation, and again the penalty may be doubled after the first

---

75/ CAA § 113(c)(5)(B), 42 U.S.C. § 7413(c)(5)(B).

76/ CAA § 113(h), 42 U.S.C. § 7413(h).

77/ CAA § 113(c)(1), 42 U.S.C. § 7413(c)(1).

78/ CAA § 113(c)(2), 42 U.S.C. § 7413(c)(2).

79/ CAA § 113(c)(4), 42 U.S.C. § 7416(c)(4).

conviction.[80]

2.6.6 EPA Settlements and Consent Orders

EPA also has authority to enter into consent decrees or consent orders to bring facilities into compliance. These decrees or orders generally include provisions for payment of civil penalties, procedures and schedules for future compliance, and stipulated penalties for any failure to comply with the requirements of the settlement. With some exceptions, the Clean Air Act requires public notice and an opportunity for public comment before EPA settlements may be finalized and submitted to a court for approval.[81]

2.6.7 EPA Penalty Policies

EPA has adopted several policies for the civil penalties it will ordinarily impose for various types of violations. Under its policy for stationary source violations, EPA employs a two-pronged approach: first, the agency will require a penalty to compensate for the full economic benefit of noncompliance plus a gravity component to reflect the seriousness of the violation; and second, EPA will consider several adjustment factors to produce what it considers a fair result.[82]

EPA's penalty policy includes methodologies for calculating the economic benefit received by the violator, including benefits from delayed or avoided costs of compliance. For example, failure to install appropriate air pollution control equipment when required benefits the violator by delaying the capital expenditure, avoiding repairs and maintenance, and paying associated operating costs. EPA generally will refuse to adjust or mitigate the economic benefit penalty to avoid creating incentives for people to await state or federal enforcement action before

---

80/ CAA § 113(c)(5), 42 U.S.C. § 7413(c)(5).

81/ CAA § 113(g), 42 U.S.C. § 7413(g).

82/ Environmental Protection Agency Penalty Policy for Stationary Source Clean Air Act Violations (January 17, 1992).

complying with applicable requirements. Under its 1992 policy statement, EPA generally will mitigate the economic benefit of noncompliance only when the benefit amount is insignificant, there are compelling public concerns that would not be served by taking the case to trial, or, in limited circumstances, there is a concurrent administrative action.

In addition to the economic benefit component of EPA penalties, the agency's policy also generally requires a gravity component designed to reflect the seriousness of the violation. Relevant factors for this gravity component include the size and economic strength of the violator, the duration and seriousness of the violation, the amount and toxicity of pollutant, the sensitivity of the affected environment, the actual or potential harm resulting from noncompliance, and the significance of the regulatory requirements at issue. Moreover, the gravity component may be increased by as much as 100 percent based on factors such as the degree of willfulness or negligence of the violator, history of noncompliance, and the extent of environmental damage. These same factors may be used to mitigate the penalty. Other mitigating factors EPA can consider include ability to pay, offsetting penalties paid to state and local governments or citizen groups for the same violations, cooperation with the government, proper reporting or correction of the noncompliance, any resulting damage, and litigation risks.[83]

2.6.8 EPA Emergency Powers

In addition to its other enforcement powers, EPA is authorized to take emergency action against any violation that presents an "imminent and substantial endangerment to public health or welfare, or the environment . . .."[84] Emergency remedies available to the agency include the issuance of administrative orders and court actions for injunctive relief. EPA has used its emergency powers sparingly.

83/ Id. at pp. 16-19.

84/ CAA § 303, 42 U.S.C. § 7603.

These emergency provisions nonetheless eliminate many of the strict notice and other procedural requirements that apply to ordinary enforcement actions under the Act and enable EPA to act promptly in emergency situations.

2.7 The Iowa Enforcement Program

Like its federal counterpart, Iowa's air pollution control program provides the DNR with a wide variety of enforcement tools. The DNR has relied heavily on notices of violations as its primary mechanism for encouraging voluntary compliance. The agency, however, has utilized other enforcement authorities to impose substantial monetary penalties on major and repeat violators.

2.7.1 DNR Notice of Violation and Revocation

DNR may issue notices of violation (NOV) to facilities not complying with their permits or state or federal air pollution control standards. An NOV usually requires correction of the violation within a specified time and may identify specific corrective actions to be taken. Typically, an NOV requires the violator to advise the agency within 30 days of receipt of notice of the facility's intended response. As discussed in section 1.2.3 of this handbook, the DNR may issue administrative penalties not to exceed $10,000 for violations of air quality statutes, rules, permits, or orders.[85]

Violation of ambient air quality standards also constitutes ground for modification or revocation of a permit, for action by the DNR to amend a stipulation agreement, or for other enforcement action by the agency to further require reduction or control of the violator's emissions.[86] A permit also may be revoked or modified based on certain other circumstances. Before the DNR may act to revoke a permit, generally it must notify the permittee of the intended revocation

---

85/ Iowa Code § 455B.109; Iowa Admin. Code § 567-10.3.

86/ Iowa Admin. Code § 567-21.4.

and its right to request a contested case before the Environmental Protection Commission.[87/] Environmental Protection Commission decisions may be reviewed in district court pursuant to the Iowa Administrative Procedure Act.[88/]

2.7.2 DNR Civil Judicial Enforcement

If action to prevent, control, or abate air pollution is not taken in response to DNR orders or permits, the DNR may ask the Iowa attorney general to commence a legal action in the name of the state for an injunction to prevent any further or continued violation.[89/] The DNR may commence civil enforcement action against a person violating any air pollution control order, permit, or rule. These enforcement actions may include a civil lawsuit seeking monetary penalties or injunctive relief, or an action to compel performance of any acts reasonably necessary to achieve compliance.[90/] The maximum civil monetary penalty the DNR may seek for air quality violations is $10,000 per day per violation.[91/]

2.7.3 DNR Criminal Enforcement

Iowa law, like its federal counterpart, has created significant criminal sanctions for violations of air pollution control requirements. A person who knowingly violates an air pollution control permit, rule, standard, or DNR order is guilty of an aggravated misdemeanor punishable by a fine of not more than $10,000 for each day of violation, two years imprisonment, or both.[92/] A person who knowingly makes a false statement, representation, or certification or knowingly fails to submit a required notification or report is guilty of an aggravated

---

87/ Iowa Code § 455B.138.

88/ Iowa Code § 455B.140; see also Iowa Code ch. 17A.

89/ Iowa Code § 455B.141.

90/ Iowa Code § 455B.146.

91/ Iowa Code § 455B.146.

92/ Iowa Code § 455B.146A(1).

misdemeanor punishable by a fine of not more than $10,000 for each day of violation, two years imprisonment, or both. A person who knowingly fails to pay a fee owed under air pollution control requirements is guilty of an aggravated misdemeanor punishable by a fine of not more than $10,000 for each day of violation, two years imprisonment, or both.

A negligent release in the air of any hazardous air pollutant or extremely hazardous substance that places another person in danger may be punished by a fine of not more than $20,000 for each day of violation, one year imprisonment, or both.[93/] A knowing release under the same conditions may be punished by a fine of not more than $50,000 for each day of violation, two years imprisonment, or both for an individual or government entity or $1,000,000 for defendants other than an individual or government entity. The penalty may be increased after the first offense.

In determining whether a defendant who is an individual "knew" that the violation placed another person in imminent danger of death or serious bodily injury, the defendant has knowledge "only if the defendant possessed actual awareness or held an actual belief."[94/]

2.7.4 DNR Emergency Powers

In addition to its other enforcement powers, the DNR is authorized to take emergency action against any violation or air pollution that presents an "emergency requiring immediate action to protect the public health and safety or property."[95/] The DNR may issue an emergency order requiring a person to reduce or discontinue the emission of air contaminants. An emergency order may be appealed to the Environmental Protection Commission.

93/ Iowa Code § 455B.146A(3).

94/ Iowa Code § 455B.146A(4)(b).

95/ Iowa Code § 455B.139.

2.8 Enforcement Tools – Recordkeeping, Reports, Inspections, Certifications, and Subpoenas

The DNR and EPA possess a variety of inspection, monitoring, and reporting powers to monitor compliance and to facilitate enforcement actions. They may require facilities to maintain records concerning operations, to submit reports, to install and use monitoring equipment, to conduct sampling, and to provide any other information that the agencies may reasonably require.96/ The agencies also may enter the premises of any source to inspect monitoring equipment or records or to sample emissions.97/

In addition, under the 1990 Clean Air Act Amendments, major stationary sources are required to submit "compliance certifications" that certify their compliance with applicable requirements.98/ Such certifications must identify, among other things, the applicable requirements, the facility's compliance status, and the method used for determining that status.

The EPA also is authorized to issue administrative subpoenas requiring the witnesses to attend proceedings and produce documents.99/ The EPA is authorized to issue such subpoenas both in connection with its investigatory and monitoring functions and for use in administrative hearings.

2.9 Citizen Suits

Private citizens are authorized to seek civil penalties in citizen suits for violations of many provisions of the Clean Air Act. The 1990 amendments created a "bounty" provision authorizing payment of up to $10,000 to anyone who provides

---

96/ CAA §114(a)(1), 42 U.S.C. § 7414(a)(1).

97/ CAA §114(a)(2), 42 U.S.C. § 7414(a)(2).

98/ CAA §114(a)(3), 42 U.S.C. § 7414(a)(3).

99/ CAA § 307(a), 42 U.S.C. § 7607(a).

information that leads to an administrative, civil, or criminal penalty.[100/] These new provisions create a potential for a significant increase in citizen suits similar to the experience under the Clean Water Act during the 1980s. Under Iowa's citizen suit statute, any person adversely affected may commence a civil action against any person for violating Iowa's air protection statutes and rules.[101/]

2.10 Chlorofluorocarbons

The federal Clean Air Act Amendments of 1990 adopted a number of specific provisions relating to stratospheric ozone protection that govern chlorofluorocarbons (CFCs). Under the federal law, certain CFCs are designated as class I substances and certain hydrochlorofluorocarbons (HCFCs) are designated as class II substances based on their ozone depletion and global warming potential.[102/] Any person who produces, imports, or exports a class I or a class II substance must monitor and file a report with the EPA setting forth the amount of the substance that was produced, imported, or exported.[103/]

The federal statute also establishes baseline years for production of class I or class II substances and requires the phase-out of production and consumption of class I substances over a period between 1991 and 2000.[104/] After January 1, 2000 (January 1, 2002 for methyl chloroform), it shall be unlawful for any person to produce any amount of a class I substance. Certain exceptions exist for essential uses of methyl chloroform, medical devices, aviation safety, national security, and for the production of limited quantities of halon for fire safety or explosion

---

100/ CAA § 113(f), 42 U.S.C. § 7413(f); CAA § 304, 42 U.S.C. § 7604; see also Save Our Health v. Recomp of Minnesota, Inc., No. 93-3284 (8th Cir. Oct. 14, 1994) (regarding Clean Air Act citizen suit notice requirements).

101/ Iowa Code § 455B.111.

102/ 42 U.S.C. § 7671a.

103/ 42 U.S.C. § 7671b. Products containing and manufactured with class I and class II substances are subject to certain labeling requirements. See 42 U.S.C. § 7671j.

104/ 42 U.S.C. § 7671c.

prevention.[105] Production and consumption of class II substances are phased out (subject to similar exceptions) effective January 1, 2030.[106]

2.11 Indoor Air -- Radon and Formaldehyde

EPA now ranks indoor air pollution as one of the four leading human health risks and as a greater danger than toxic outdoor air pollution, carbon monoxide, hazardous waste, or contaminated groundwater. Studies indicate that indoor levels of airborne contaminants -- such as formaldehyde, radon, asbestos,[107] biological contaminants (like bacteria, viruses, molds, and spores), and secondhand tobacco smoke -- may be much higher than outdoor levels. Despite the concern regarding indoor air pollution, there is currently no comprehensive federal or Iowa law aimed to protect indoor air quality. Congress is considering legislation to establish standard methods for measuring and identifying indoor air emissions, set standards of indoor air quality, require labeling of products that emit pollutants, and require that new or renovated buildings meet minimum ventilation standards.[108]

An indoor air contaminant that has received considerable attention is radon. Radon is a colorless, odorless, tasteless, radioactive gas that occurs naturally in soil, groundwater, rock, and outdoor air. It exists at various levels throughout the United States. Prolonged exposure to elevated levels of radon has been reportedly associated with increases in the risk of lung cancer.

Under the federal Clean Air Act, the EPA has adopted National Emission Standards for Hazardous Air Pollutants (NESHAPs) that govern radon emissions from three kinds of facilities.[109] Specifically, these radon emission standards apply

---

105/ 42 U.S.C. § 7671c(d) and (g).

106/ 42 U.S.C. § 7671d(b)(2).

107/ Asbestos regulation is discussed in chapter 13 of this handbook.

108/ OSHA has proposed controversial indoor air quality standards. 59 Fed. Reg. 15,968 (April 5, 1994).

109/ 42 U.S.C. § 7412.

to Department of Energy facilities, phosphogypsum stacks, and uranium mill tailings.110/ For all three types of facilities, emissions of radon-222 must not exceed 20 picoCuries per cubic meter (pCi/m^3).

In 1988, Congress mandated that the "national long-term goal of the United States with respect to radon levels in buildings is that the air within buildings in the United States should be as free of radon as the ambient air outside of buildings."111/ Currently, the EPA recommends that indoor levels of radon be reduced to less than 4 picoCuries per liter (pCi/l). Although exposures below this level may present some risk of lung cancer, reductions to levels below 4 pCi/l may be difficult, and sometimes impossible to achieve.112/ Information on obtaining a radon detector can be obtained from the American Lung Association.

The Iowa Department of Public Health has established a program to certify persons who test for radon.113/ Although stringent confidentiality provisions govern this program, certified testing services must disclose to the Department of Public Health the address and test results if EPA guidelines are exceeded.114/

Another common concern regarding indoor air is formaldehyde. The U.S. Department of Housing and Urban Development (HUD) has established product standards, certification, and labeling requirements for urea formaldehyde-containing particle board and plywood used in manufactured housing, which are generally intended to insure that indoor air levels of formaldehyde do not exceed 0.4 parts per million.

Persons injured by exposure to indoor air pollutants have asserted claims under various common law doctrines, including breach of contract, breach of warranty, strict liability, negligence, fraud or misrepresentation, nuisance, assault,

---

110/ 40 C.F.R. §§ 61.190-61.225.

111/ 15 U.S.C. § 2661.

112/ EPA, A Citizen's Guide to Radon (1986).

113/ Iowa Code § 136B.1; Iowa Admin. Code ch. 641-44.

114/ Iowa Code § 136B.2.

and infliction of emotional distress. Some plaintiffs also have recovered cleanup costs associated with indoor air pollutants under Superfund laws.[115]

2.12 Iowa Clean Indoor Air Act -- Smoking

The Iowa Clean Indoor Air Act prohibits smoking in public places and public meetings except in designated smoking areas.[116] Smoking is defined as "carrying of or control over a lighted cigar, cigarette, pipe, or any other lighted smoking equipment."[117] The term "public place" means "any enclosed indoor area used by the general public or serving as a place of work containing 250 or more square feet of floor space, including, but not limited to, all restaurants with a seating capacity greater than 50, all retail stores, lobbies, and malls, offices including waiting rooms, and other commercial establishments; public conveyances with departures, travel, and destination entirely within this state; educational facilities; hospitals, clinics, nursing homes, and other health care and medical facilities; and auditoriums, elevators, theaters, libraries, art museums, concert halls, indoor arenas, and meeting rooms."[118] Private, enclosed offices occupied exclusively by smokers even though such offices may be visited by non-smokers are excluded.

Persons in charge of a public place may designate smoking areas, except where smoking is prohibited by the fire marshal or by other law.[119] Persons in charge of a public place also must take reasonable efforts to prevent smoking in the

---

115/ See Bloomquist v. Wapello County, 500 N.W.2d 1 (Iowa 1993); see also Vermont v. Staco, Inc., 684 F. Supp. 822 (D. Vt. 1988); BCW Associates, Inc. v. Occidental Chem. Corp., 1988 Westlaw 102641 (E.D. Pa. 1988); Cyker v. Four Seasons Hotel, Ltd., 1991 Westlaw 1401 (D. Mass. 1991). Common law causes of action are discussed in chapter 14 of this handbook.

116/ Iowa Code § 142B.2; see also 59 Fed. Reg. 15,968 (April 5, 1994) (OSHA proposed indoor air quality standards).

117/ Iowa Code § 142B.1(4).

118/ Iowa Code § 142B.1(3).

119/ Iowa Code § 142B.2(2).

public place.[120] Any person who violates the smoking prohibition may be subject to a civil penalty of $25.[121]

2.13 Pollution Standards Index

The EPA requires all state air pollution agencies to report an air pollution standards index (PSI) for cities with a population of 200,000 or more. This index is based on the composite value of major air pollutants such as total particulates, ozone, carbon monoxide (CO), and sulfur dioxide (SO_2). The reported index value is based on the pollutant concentration level which, during the previous 24-hour period, presents the greatest threat to human health. During the hay fever season, mold and pollen counts are also included in the index. The PSI values range from 0 to 500, according to the magnitude of their health effects. A PSI value of 100 corresponds to pollutant concentrations that equal the short-term (24 hours or less) federal primary standards. The PSI is reported in the following manner:

PSI Value	Air Quality	Warning
0 - 49	GOOD	
50 - 99	MODERATE	
100 - 199	UNHEALTHFUL	Persons with existing heart or respiratory ailments should reduce physical exertion and outdoor activity.
200 - 299	VERY UNHEALTHFUL	Elderly persons, and persons with existing heart disease, should stay indoors and reduce physical activity.
300 - 399	HAZARDOUS	Elderly persons and persons with existing diseases should stay indoors and avoid physical exertion; general public should avoid outdoor activity.
400 - 499	HAZARDOUS	All persons should remain indoors keeping windows and doors closed; all persons should minimize physical exertion and avoid traffic.

120/ Iowa Code § 142B.3.

121/ Iowa Code § 142B.6; see also Iowa Code § 805.8(11).

CHAPTER THREE
WATER POLLUTION CONTROL

3.0 Introduction To Water Pollution Control Regulation

Iowa's water resource regulations govern all of the following:

- surface water quality;
- the treatment of pollution discharged into water;
- the construction and operation of water treatment facilities;
- runoff from agricultural facilities;
- runoff of storm waters from industrial and construction sites; and
- the quality of groundwater.

Iowa's program for regulating water pollution[1] authorizes the Environmental Protection Commission (EPC), acting under the umbrella of the DNR, to "[d]evelop comprehensive plans and programs for the prevention, control, and abatement of water pollution."[2] The EPC's policy statement adds that "it will attempt to prevent and abate the pollution of all waters to the fullest extent possible consistent with statutory, and technological limitations. This policy is to apply to all point and nonpoint sources of pollution."[3] Pursuant to this policy, the EPC sets water quality standards for the surface waters of the state. The state may enforce these standards through civil and criminal penalties.[4] This chapter provides a brief overview of some of the primary laws and regulations of Iowa's water regulatory program.

3.1 General Considerations on Water Quality Control

Iowa controls discharge of water pollution according to both water quality and

---

1/ See generally Iowa Code ch. 455B; Iowa Admin. Code ch. 567-60 to 69.

2/ Iowa Code § 455B.173.

3/ Iowa Admin. Code § 567-61.2(1).

4/ See Iowa Code § 455B.191.

minimum effluent treatment standards.5/ Pursuant to its water quality standards, Iowa classifies bodies of water according to the water quality the waters' "use" requires.6/ For instance, a body of water that provides drinking water demands more stringent protections than water of more limited use to humans or aquatic life.

Under Iowa's anti-degradation policy, discharge of pollutants must not cause a body of water to exceed the pollutant level allowed for that water's use.7/ It also requires that, if water quality significantly exceeds the quality necessary for a body of water's use classification, the state will preserve the water's high quality levels unless, after public hearings, the state determines that "necessary and justifiable economic and social developments in the area" require that the state permit lower chemical quality.8/

Iowa also regulates, through the use of minimum effluent standards, discharges of pollutants that would not cause the receiving water to exceed water quality pollutant criteria. All point source9/ waste waters must "receive treatment in compliance with minimum effluent standards."10/

5/ "'Effluent standard' means any restriction or prohibition on quantities, rates, and concentrations of chemical, physical, biological, radiological and other constituents which are discharged from point sources into any water of the state including an effluent limitation, a water quality related effluent limitation, a standard of performance for a new source, a toxic effluent standard or other limitation." Iowa Code § 455B.171(5).

6/ Iowa Admin. Code § 567-61.3.

7/ Iowa Admin. Code § 567-61.2(2); see also IBP, Inc. v. Iowa DNR, 515 N.W.2d 554 (Iowa App. 1994).

8/ Iowa Admin. Code § 567-61.2(2).

9/ "'Point Source' means any discernible, confined, and discrete conveyance, including but not limited to any pipe, ditch, channel, tunnel, conduit, well, discrete fissure, container, rolling stock, concentrated animal feeding operation, or vessel or other floating craft, from which pollutants are or may be discharged." Iowa Code § 455B.171(12). "Non-point source" is a term commonly used to refer to other sources of runoff.

10/ Iowa Admin. Code § 567-61.2(3).

3.2 NPDES Permits

Both federal and state requirements apply to the direct discharge of pollutants into state waters. Iowa, however, has received authorization from the United States Environmental Protection Agency (EPA) to administer the federal standards that have been established under the Clean Water Act, including the National Pollutant Discharge Elimination System (NPDES) program.[11/] The Iowa NPDES program consists of these federal standards, as well as additional state effluent limitations and water quality standards.

Iowa's NPDES program establishes and enforces effluent limitations and water quality standards through a permit system designed to control the discharge of pollution.[12/] In accordance with Iowa's EPA authorization, an NPDES permit or operation[13/] permit is required to discharge a pollutant from a point source into navigable waters.[14/] The term "pollutant" is defined by statute very broadly to include "sewage, industrial waste or other waste."[15/]

11/ See 33 U.S.C. § 1342(b); 40 C.F.R. pts. 123, 125.

12/ See Iowa Admin. Code ch. 567-64.

13/ NPDES permits are a form of operation permit. " 'Operation permit' means a written permit by the [DNR] director authorizing the operation of a wastewater disposal system or part thereof or discharge source, and, if applicable, the discharge of wastes from said disposal system or part thereof or discharge source to waters of the state." Iowa Admin. Code § 567-64.1(4). Operation permits should be distinguished from the "construction permits" required to construct a wastewater disposal system or a part of such a system. Iowa Admin. Code § 567-64.1(2).

14/ Iowa Admin. Code § 567-62.1(1); see also Iowa Admin. Code § 567-60.2 (" 'Navigable water' means a water of the United States.").

15/ Iowa Code § 455B.171(13). "Sewage" means "water-carried waste products from residences, public buildings, institutions, or other buildings, including the bodily discharges from human beings or animals together with such groundwater infiltration and surface water as may be present." Iowa Code § 455B.171(21). "Industrial waste" means "any liquid, gaseous, radioactive, or solid waste substance resulting from any process of industry, manufacturing, trade or business or from the development of any natural resource." Iowa Code § 455B.171(7). "Other waste" means "heat, garbage, municipal refuse, lime, sand, ashes, offal, oil, tar, chemicals and all other wastes which are not sewage or industrial waste." Iowa Code § 455B.171(10).

3.2.1 NPDES Exemptions and Variances

The following forms of discharge do not require an operation permit:

1. Private sewage disposal systems not discharging into a water of the state;[16]

2. Semi-public sewage disposal systems constructed with departmental approval and which do not discharge into a water of the state;

3. Discharges of sewage from vessels, effluent from properly functioning marine engines, laundry, shower, and galley sink wastes, or any other discharge incidental to the normal operation of a vessel but excluding discharges of garbage overboard or discharges of vessels not used as a means of transportation;

4. Discharges of aquacultural projects as defined by 40 C.F.R. § 122.25;

5. Discharges of pollutants directly into another waste disposal system for final treatment and disposal but excluding discharge of storm water point sources;

6. Discharges in compliance with instructions of an On-Scene Coordinator pursuant to 40 C.F.R. Part 300 or 33 C.F.R. § 153.10(e);

7. Water pollution stemming from agricultural and silvicultural activities, orchard runoff, cultivated crops, pastures, range-lands, and forest lands, but not including:

    a. Discharges from concentrated aquatic animal production facilities as defined by 40 C.F.R. § 122.27;

    b. Discharges from concentrated animal feeding operations as defined by 40 C.F.R. § 122.23;

    c. Discharges from silvicultural point sources as defined in 40 C.F.R. § 122.27;

16/ " 'Water of the state' means any stream, lake, pond, marsh, watercourse, waterway, well, spring, reservoir, aquifer, irrigation system, drainage system, and any other body or accumulation of water, surface or underground, natural or artificial, public or private, which are contained within, flow through or border upon the state or any portion thereof." Iowa Code § 455B.171(26); see also State v. Butler, 419 N.W.2d 362 (Iowa 1988) (upholding local health board's authority over sewage disposal pursuant to Iowa Code § 137.21).

d. Storm water discharges associated with industrial activity as defined by Iowa Admin. Code § 567-60; and

8. Return flows from irrigated agriculture.[17]

In addition, the DNR director may grant exemptions to dischargers from requirements on maximum contaminant levels or minimum treatment techniques.[18] If however, the discharger seeks a variance from federal effluent limitations that Iowa has adopted by reference, then the discharger must demonstrate to the DNR director that factors relating to the discharger's equipment, facilities, process, or other factors related to the discharger are "fundamentally different" from the factors the EPA administrator considered in developing effluent limitations.[19] The DNR director must then make a written finding whether the discharger's relevant factors meet this requirement. Furthermore, the EPA administrator must, as a "condition precedent," approve any such variance from the federal standards.

3.2.2 NPDES Individual Permit Application Process

An applicant must generally file an application for an individual NPDES permit at least 180 days prior to the date operation is scheduled to begin,[20] unless the DNR director approves a shorter period.[21] Applications for individual permits must be prepared on forms provided by the DNR and must be accompanied by the

---

17/ For precise text of exceptions, see Iowa Admin. Code § 567-64.3(1).

18/ See Iowa Code § 455B.181.

19/ Iowa Admin. Code § 567-62.7.

20/ Individual permit applications for discharge of storm water associated with construction activity, as defined in the Iowa Admin. Code § 567-64.3(4), are due only 60 days prior to the date on which construction is to begin. Iowa Admin. Code § 567-64.3(4)(a).

21/ Iowa Admin. Code § 567- 64.3(4).

appropriate permit fee.22/ The director has discretion to require additional information as necessary to evaluate the application.

The DNR first makes a tentative determination of whether to grant an NPDES permit. If the tentative determination is to grant the permit, the DNR then prepares a draft permit that contains: (1) effluent limitations; (2) if necessary, a proposed schedule for compliance; and (3) special conditions, "which will have a significant impact on the discharge described in the NPDES application."23/

Before issuing an NPDES permit, DNR must give public notice of the proposed permit and provide a minimum of 30 days following the date of public notice for interested persons to submit written comments on the DNR's tentative NPDES determination. It must also provide notification to appropriate governmental agencies of completed NPDES applications.24/ The DNR must retain all written comments, and the DNR director must consider them when making a final decision as to whether to grant the NPDES permit.25/

Each public notice of a proposed NPDES permit should contain: (1) the name, address and phone number of the DNR; (2) the name and address of each applicant; (3) a brief description of the applicant's processes that result in discharge; (4) the name of the receiving waterway, the location of each discharge on the waterway, and indication whether the discharge is new or existing; (5) a statement of the DNR's initial determination as to whether to grant or deny the permit; (6) a brief description of the procedures used for final determination; and (7) the address and phone number of places from which interested persons may obtain further

---

22/ Permit fees are specified in Iowa Admin. Code § 567-64.16.

23/ Iowa Admin. Code § 567-64.5(1).

24/ Iowa Admin. Code § 567-65.5(4).

25/ Iowa Admin. Code § 567-64.5(2)(a,b).

information or request copies of fact sheets,26/ the draft permit, NPDES forms, and related documents.27/

Interested parties may request a public hearing on the NPDES permit within the prescribed 30-day comment period. The request must state the interest of the filing party and the reasons that a hearing is necessary. The director shall hold a public hearing if significant public interest warrants. Public notice must be provided for public hearings in accord with DNR policies.28/

3.2.3 General Permit Application Process -- Storm Water

Iowa administers a general permitting process for certain NPDES permits -- most notably, for storm water discharges. General permits, unlike individual permits, are not tailored to individual dischargers. Rather, they specify terms and conditions which dischargers must meet to be eligible for the general permit. Applicants must generally file for a general permit "at least 24 hours prior to the date operation is scheduled to begin."29/

To obtain a general permit, an applicant must file a Notice of Intent for coverage on the appropriate DNR forms. Different forms are required for "Storm Water Discharge Associated with Industrial Activity"30/ and "Storm Water

---

26/ Fact sheets must contain a description of the location of discharge, a quantitative description of discharge (i.e., average daily flow, average temperatures for thermal pollution, average daily discharge in pounds of significant pollutants, etc.), water quality standards of the receiving water, minimum effluent treatment standards for discharge, tentative DNR determinations on the grant of an NPDES permit, and a full description of procedures used to reach a final determination. Iowa Admin. Code § 567-64.5(3)(a). The DNR must prepare a fact sheet for every discharge which exceeds 500,000 gallons on any day of the year. It must make fact sheets relating to a discharge described in a public notice available on request.

27/ Iowa Admin. Code § 567-64.5(2)(c).

28/ Iowa Admin. Code § 567-64.5(6).

29/ Iowa Admin. Code § 567-64.3(4)(b)(3).

30/ For a definition of "storm water associated with industrial activities" see Iowa Admin. Code § 567-60.2.

Discharge Associated with Industrial Activity for Construction Activities."31/ Both forms, however, require the following:

1. The name, mailing address, and location of the discharging facility and its owner;
2. The four-digit SIC (Standard Industrial Classification) code "that best represents the principal products or activities provided by the facility;"
3. The operator's name, address, telephone number, ownership status and status as federal, state, private, public or other entity;
4. The discharging facility's township, range and county, 1/4 section, or its latitude and longitude;
5. The discharge type, date of commencement of discharge, permit status of discharge, whether or not discharge is to a municipal storm sewer, and the name of the receiving water; and
6. Available quantitative data describing the concentration of pollutants in the storm water discharge.32/

In addition, construction sites requiring a storm water discharge permit must provide a brief description of the construction project, a timetable for major activities, and an estimate of the number of acres of soil that will be disturbed.33/

Applications must include the appropriate permit fee. Applications also must demonstrate that the applicant has provided public notice of the proposed discharge newspaper publication and that such notice met the minimum informational requirements mandated by the DNR.34/ Further, applicants must

---

31/ Construction activities that require a storm water discharge permit include, "clearing, grading and excavation activities except: operations that result in the disturbance of less than five acres of total land area which are not part of a larger common plan of development or sale." 40 C.F.R. § 122.26(b)(14)(x) (adopted by reference by Iowa Admin. Code § 567-64.13.)

32/ Iowa Admin. Code § 567-64.6(1)(a).

33/ Iowa Admin. Code § 567-64.6(1)(a)(7).

34/ Iowa Admin. Code § 567-64.6(1)(c).

develop a storm water pollution prevention plan designed to prevent stormwater from carrying pollutants off the site.

Upon submission of a complete application package in accord with DNR rules, the applicant is authorized to discharge unless otherwise notified by the DNR.[35/] A permit-holder intending to discontinue the authorized discharge should file a notice of discontinuance within 30 days prior to or after discontinuance. Such notice must contain: (1) the name of the discharging facility; (2) the general permit number and permit authorization number; (3) the date of discontinuance; and (4) "a signed certification in accordance with the requirements in the general permit."[36/]

3.2.4 Conditions of NPDES Permits

Each NPDES permit states the various legal requirements that the discharger must meet to obtain NPDES authorization. In addition, each permit must include a statement from the DNR director that the authorized discharge does not cause violation of applicable standards.[37/] Applying these standards, the DNR director must "specify average and maximum daily quantitative limitations for the level of pollutants in the authorized discharge."[38/]

Iowa regulations provide a set of guidelines for the DNR director to use in setting compliance schedules for dischargers.[39/] In case of noncompliance, the permittee shall be required to take "specific steps" to reach compliance. If there is no specified compliance schedule, the permittee must take such steps in the "shortest, reasonable period of time. . . consistent with the guidelines and requirements" of

---

35/ Iowa Admin. Code § 567-64.6(2).

36/ Iowa Admin. Code § 567-64.6(5).

37/ See Iowa Admin. Code § 567-64.7(2).

38/ Iowa Admin. Code § 567-64.7(3); see also IBP, Inc. v. Iowa DNR, 515 N.W.2d 554 (Iowa App. 1994) (regarding mixing zones to determine effluent limits).

39/ Iowa Admin. Code § 567-64.7(4).

the Clean Water Act.[40] In cases in which the schedule of compliance would exceed nine months, the permit must set dates for achievement of interim requirements.[41] Failure or refusal to meet interim or final requirements is a violation which can justify modification, suspension, or revocation of the NPDES permit or direct enforcement action.[42]

The Iowa Admin. Code lists numerous other conditions for NPDES permits.[43] Notable conditions include: (1) reporting requirements for "facility expansions, production increases, or process modifications which result in new or increased discharges of pollutants;" (2) a requirement that if the terms of a general permit no longer apply to discharge, the general permit holder will seek an individual permit;[44] and (3) requirements that the permittee allow the DNR director or the director's authorized representatives to inspect and copy required records, to inspect monitoring equipment and methods, and to sample any discharge of pollutants.[45]

3.2.5 Term, Revocation, and Modification of NPDES Permit

The term of an NPDES operation permit may not exceed five years.[46] The DNR director, for cause, may modify, suspend, or revoke, in whole or in part, individual operation or general permits.[47] Iowa Admin. Code § 567-64.3(11) states that cause includes:

a. Violation of any term or condition of the permit;

40/ Iowa Admin. Code § 567-64.7(4)(a).

41/ Iowa Admin. Code § 567-64.7(4)(b).

42/ Iowa Admin. Code § 567-64.7(4)(e); see also section 3.11 of this handbook.

43/ See generally Iowa Admin. Code § 567-64.7(5) (listing various conditions for NPDES permits).

44/ Iowa Admin. Code § 567-64.7(5)(a).

45/ Iowa Admin. Code § 567-64.7(5)(c).

46/ Iowa Admin. Code § 567-64.3(7).

47/ Iowa Admin. Code § 567-64.3(11).

b. Obtaining a permit by misrepresentation of fact or failure to disclose fully all material facts;

c. A change in any condition that requires either a temporary or permanent reduction or elimination of the permitted discharge;

d. Failure to submit such records and information as the director shall require both generally and as a condition of the operation permit in order to assure compliance with the discharge conditions specified in the permit;

e. Failure or refusal of an NPDES permittee to carry out the requirements of Iowa Admin. Code § 567-64.7(5)(c) [stating the DNR's authorization to inspect and copy records, inspect monitoring equipment and methods, and sample discharge];

f. Failure to provide all the required application materials or appropriate fees.48/

In addition, the director may suspend general permit coverage and require a permittee to apply for an individual permit when either: (1) the discharge would cause violation of water quality standards; or (2) the DNR determines that the permitee's activities violate the conditions of the general permit.49/

3.3 Animal Feeding Operations

3.3.1 Animal Feeding Operations Requiring a Permit

Iowa requires permits for the operation and construction of certain animal-feeding operations.50/ An operation permit is generally required for: (1) open

---

48/ Iowa Admin. Code § 567-64.3(11).

49/ Iowa Admin. Code § 567-64.6(3).

50/ " 'Animal feeding operation' means a lot, yard, corral, building, or other area in which animals are confined and fed and maintained for forty-five days or more in any twelve-month period. Two or more animal feeding operations under common ownership or management are deemed to be a single animal feeding operation if they are adjacent or utilize a common area or system for waste disposal." Iowa Admin. Code § 567-65.1; see also DeCoster vo Franklin County, 497 N.W.2d 849 (Iowa 1993) (DNR, but not Franklin County, had jurisdiction to regulate construction of livestock waste storage basin); Concerned Area Residents v. Southview Farm, 1994 Westlaw 480646 (2d Cir. 1994) (dairy farm is "point source" under Clean Water Act).

feedlots51/ exceeding specified capacities that vary with the animal or fowl;52/ or (2) operations that may not exceed the specified capacity, but which discharge wastes either directly or through a human-made waste drainage system "into a water of the state which originates outside of and traverses the operation."53/

The DNR may evaluate, on an individual basis, any animal feeding operation to determine whether the operation is discharging waste into waters of the state and failing to achieve minimum levels of waste control or is "causing or may be reasonably expected to cause pollution of the waters of the state" or "a violation of state water quality standards."54/ If an operation fails any of these criteria, the DNR may require the operation to obtain a permit or take remedial action.55/

In addition to the operation permit, an animal feeding operation of sufficient size to trigger operation permit requirements must "obtain a construction permit prior to constructing, installing, or modifying the waste control system for that operation."56/ Confinement feeding operations57/ that use certain waste control methods and/or exceed a certain size must obtain construction permits.58/

51/ "'Open feedlot' means an unroofed or partially roofed animal feeding operation in which no crop, vegetation, or forage growth or residue cover is maintained during the period that animals are confined in the operation." Iowa Admin. Code § 567-65.1.

52/ See Iowa Admin. Code § 567-65.3(1) (detailing open feedlot capacities that trigger permit requirements, e.g. 1000 beef cattle, 700 dairy cattle, 2,500 swine, 55,000 turkey).

53/ See Iowa Admin. Code § 567-65.3(2) (detailing capacity requirements for such operations that trigger permit requirements, e.g. 300 beef cattle, 200 dairy cattle, 750 swine, 16,500 turkeys).

54/ Iowa Admin. Code § 567-65.4(1).

55/ Iowa Admin. Code § 567-65.4(2). The DNR may not, however, pursuant to its evaluation authority, require an operation with an animal capacity below that set forth in Iowa Admin. Code § 567-65.3(2) to obtain an operation permit unless the operation discharges directly or through a man-made waste drainage system into the waters of the state.

56/ Iowa Admin. Code § 567-65.6(1)(a).

57/ "'Confinement feeding operation' means totally roofed animal feeding operation in which wastes are stored or removed as a liquid or semiliquid." Iowa Admin. Code § 567-65.1.

58/ Iowa Admin. Code § 567-65.6(1)(b) (e.g., 200 beef cattle if earthen waste storage facility, 2,000 beef cattle if steel tank).

3.3.2 Minimum Waste Control Requirements for Animal Feedlots

Iowa provides various minimum waste control requirements applicable to animal feeding operations in general, open feedlots, and confinement feeding operations. All animal feeding operations must remove all "settleable solids from . . . wastes prior to discharge into a water of the state."59/ Open feedlots requiring an operations permit (other than a permit required due to DNR individualized evaluation) must also provide for "retention of all waste flows from the feedlot areas and all other waste-contributing areas resulting from the 25-year, 24-hour precipitation event."60/ Confinement feeding operations, by contrast, must retain "all wastes produced in the confinement enclosures between periods of waste disposal. In no case shall wastes from a confinement feeding operation be discharged directly into a water of the state or into a tile line that discharges to the waters of the state."61/ The DNR may also, as necessary to provide for adequate water pollution control, tailor minimum waste control requirements to suit the needs of specific animal feeding operations.62/

Wastes removed from an animal feeding operation should be disposed of "by land application in a manner which will not cause surface or groundwater pollution."63/ In addition, animal feeding operations must not discharge directly into a publicly owned lake, a sinkhole, or an agricultural drainage well.

3.3.3 Animal Feedlot Permit Conditions

The DNR includes in animal feedlot operation (and construction) permits the conditions it deems necessary to assure compliance with DNR rules, proper

---

59/ Iowa Admin. Code § 567-65.2(1).

60/ Iowa Admin. Code § 567-65.2(2).

61/ Iowa Admin. Code § 567-65.2(3).

62/ Iowa Admin. Code § 567-65.2(4).

63/ Iowa Admin. Code § 567-65.2(7).

operation and maintenance, protection of public health, protection of beneficial uses of state waters, and the prevention of water pollution.64/ In addition, the DNR may impose monitoring and reporting requirements on animal feeding operations.

An operations permit may not authorize more than five years of operation. To renew a permit, an applicant must apply no later than 180 days prior to expiration. Renewals will be subject to the rules of the DNR that cover the applying facility at the time of renewal application.65/

For cause, the DNR has discretion to modify, suspend, or revoke, in whole or in part, any operation permit. Iowa Admin. Code § 567-65.5(11) states that cause includes:

a. Violation of any term or condition of the permit;

b. Obtaining a permit by misrepresentation of fact or failure to disclose fully all material facts;

c. A change in any condition that requires either a temporary or permanent reduction or elimination of the permitted discharge; or

d. Failure to submit the records and information that the DNR requires in order to assure compliance with the operation and discharge conditions of the permit.66/

3.4 NPDES Monitoring

Iowa Admin. Code ch. 567-63 details minimum self-monitoring and reporting requirements that various kinds of dischargers must obey. It also allows the DNR to specify additional monitoring on a case-by-case basis. Notably, all NPDES permits contain conditions allowing the DNR director or director's authorized representatives:

64/ Iowa Admin. Code § 567-65.5(9).

65/ Iowa Admin. Code § 567-65.5(10).

66/ For further discussion of environmental laws applicable to Iowa farmers, see the sections on nuisance law in chapter 14 of this handbook and N. Hamilton, What Farmers Need To Know About Environmental Law (1990), available from Drake University.

1. To enter the premises where an effluent source is located or where records required by permit are kept;
2. To have access to and copy required records;
3. To inspect required monitoring equipment and methods; and
4. To sample discharges.[67/]

Failure to permit agency monitoring is grounds for revocation or suspension of the permit.[68/]

3.5 Notification of Unintentional By-Passes

A discharging facility must notify the DNR of by-passes of sewage or waste water that occur as a result of mechanical failure or acts beyond the control of the owner but not as a result of precipitation. Notification must come within twelve hours of by-pass. It must state the reasons for and expected duration of the by-pass. The owner must then obey any instructions from the DNR designed to minimize damage to receiving waters.[69/]

3.6 Publicly Owned Treatment Works (POTW) Requirements

Iowa Admin. Code 567-62.3 sets forth various minimum secondary[70/] treatment effluent standards that both publicly owned treatment works (POTW) and privately owned domestic sewage treatment works must meet. These standards state requirements on carbonaceous biochemical oxygen demand, suspended solids,

---

67/ Iowa Admin. Code § 567-64.7(5)(c).

68/ Iowa Admin. Code § 567-64.3(11)(e).

69/ Iowa Admin. Code § 567-63.6(2).

70/ Under certain conditions, treatment works are eligible to use "treatment equivalent to secondary treatment" rather than secondary treatment to meet effluent requirements. Iowa Admin. Code § 567-62.3(3). This rule also states the minimum effluent standards facilities using "equivalent" treatments must meet. Iowa Admin. Code § 567-62.3(3).

and pH levels.71/ Effluent limitations on other pollutants may be required by a facility's NPDES permit.

Owners of POTWs must "prepare and implement a plan of action to achieve and maintain compliance with final effluent limitations." The plan of action for a given POTW depends on its compliance history and physical status. The plan must identify "deficiencies and needs" of the POTW and their causes. Furthermore, it must propose remedial steps as well as an implementation schedule and a means of financing for such steps.72/ In carrying out these plans, POTWs in some circumstances require pretreatment of industrial wastewater before it will be accepted by the POTW.

An operator of a wastewater treatment plant must be certified by the DNR.73/ The certification is in addition to the NPDES operation permits required by Iowa Code § 455B.273. The level of competency certified depends on the population served and amount of wastewater treated.74/

The state has established a sewage work construction fund to assist in funding municipal sewer construction projects.75/ Similarly, both EPA and the state legislature provide funding for the state revolving fund for wastewater treatment plants.76/ Allocation of loan funds is based on an Intended Use Plan (IUP) prepared by the DNR. DNR regulations set forth procedures for municipal applicants for loan assistance.77/

71/ Iowa Admin. Code § 567-63.3(1).

72/ For details of POTW plan of action requirements, see Iowa Admin. Code § 567-64.7(6). For additional conditions applicable to POTWs, see Iowa Admin. Code § 567-64.7(5)(d,e).

73/ See Iowa Code §§ 455B.211-455B.224; Iowa Admin. Code ch. 567-81.

74/ Iowa Admin. Code ch. 567-81.

75/ See Iowa Code §§ 455B.291-455B.300; Iowa Admin. Code ch. 567-90 to 92.

76/ Iowa Code § 455B.293; 40 C.F.R. pts. 30, 33, 35.

77/ Iowa Code ch. 567-92.

3.7 Groundwater Protection

In addition to surface water pollution regulations, Iowa has enacted statutes and promulgated regulations to control pollution of groundwater.[78/] The goal of Iowa's groundwater program is "to prevent contamination of groundwater from point and nonpoint sources of contamination to the maximum extent practical, and if necessary to restore the groundwater to potable state, regardless of present condition, use, or characteristics."[79/] Iowa has granted the EPC authority to adopt rules to implement this policy[80/] and granted the DNR director investigative and enforcement powers.[81/] Iowa also has encouraged political subdivisions to implement their own groundwater protection plans "provided that implementation is at least as stringent but consistent with" DNR rules.[82/]

Groundwater protection rules provide mechanisms for documenting contamination, sources of contamination, and the identity of "responsible persons."[83/] "Responsible persons" are persons legally liable for contamination or responsible for abating contamination.[84/]

The required response to contamination varies with the seriousness of contamination. As an initial matter, to prevent further contamination, the rules require immediate response to control readily correctable active sources of

---

78/ See Iowa Code ch. 455E; Iowa Admin. Code chs. 567-15; 561-133. " 'Groundwater' means any water of the state, as defined in section 455B.171, which occurs beneath the surface of the earth in a saturated geological formation of rock or soil." Iowa Code § 455E.2(7).

79/ Iowa Code § 455E.4.

80/ Iowa Code § 455E.9.

81/ Iowa Code § 455E.8(8).

82/ Iowa Code § 455E.10.

83/ Iowa Admin. Code § 567-133.3.

84/ The category of responsible person "may include the person causing, allowing or otherwise participating in the activities or events which cause the contamination, persons who have failed to conduct their activities so as to prevent the release of contaminants into groundwater, property owners who are obligated to abate a condition, or persons responsible for or successor to such persons." Iowa Admin. Code § 567-133.2; see also Blue Chip Enters. v. Iowa DNR, No. 75-220-891 (Hardin County Dist. Ct. April 23, 1993).

contamination.85/ The rules also provide for immediate removal of "readily accessible contaminants . . . to avoid or minimize further contamination in the groundwater."86/

In cases where contamination poses a "significant risk,"87/ the responsible party must take steps "to determine the extent and levels of contamination. . ."88/ The responsible party must also submit a "remedial action plan."89/ The DNR may require the responsible party to conduct further investigation of contamination or to supplement its remedial action plan.90/

In cases where contamination poses an "aggravated risk,"91/ the rules provide for expedited preventative, investigatory, and remedial measures.92/ The rules also may require responsible parties to take emergency steps to protect the environment and public health, such as to provide alternate water supplies, fence off contaminated areas, or recommend evacuations to appropriate authorities. 93/

As a general matter, the rules describe the goal of cleanup as:

> . . . use of best available technology and best management practices as long as it is reasonable and practical to remove all contaminants, and in any event until water contamination remains below the action level for any

---

85/ Iowa Admin. Code § 567-133.4(1).

86/ Iowa Admin. Code § 567-133.4(1).

87/ See Iowa Admin. Code § 567-133.2 (defining "significant risk").

88/ Iowa Admin. Code § 133.4(3)(a).

89/ " 'Remedial action plan' means a written report which includes all relevant information, findings, and conclusions from a site assessment, including all analytical results and identification of contaminant migration pathways; identification and evaluation of cleanup alternatives, including both active and passive measures using best available technology and best management practices; a recommended cleanup action or combination of action, including identification of expected cleanup levels consistent with the cleanup goal of 133.4(3)(b) a monitoring network and schedule to document cleanup levels; and a proposed schedule of implementation." Iowa Admin. Code § 567-133.2.

90/ Iowa Admin. Code § 567-133.4(3)(a).

91/ See Iowa Admin. Code § 567-133.2 (defining "aggravated risk").

92/ Iowa Admin. Code § 567-133.4(2).

93/ Iowa Admin. Code § 567-133.4(2).

> contaminant, and the [DNR] determines that the contamination is not likely to increase and no longer presents a significant risk.94/

Where site or technological concerns make this goal impractical, however, the DNR can establish "alternative cleanup level or levels, including such other conditions as will adequately protect the public health, safety, environment, and quality of life."

Liability under Iowa Code ch. 455E, which is devoted specifically to groundwater protection, requires a violation of Iowa Code ch. 455B, which governs pollution control more generally.95/ Iowa Code § 455E.6 also exempts certain agricultural producers from liability for active cleanup costs and damages due to detection in groundwater of nitrates and pesticides. This exemption is conditioned on the producer's compliance, as appropriate, with soil test results, licensing requirements, and labelling instructions for application of fertilizer or pesticide.

3.8 Public Water Supply Systems -- Safe Drinking Water Act

The federal Safe Drinking Water Act and its implementing regulations establish a separate program to protect drinking water supplies.96/ While the EPA sets the minimum standards for this program, the DNR is responsible for conducting and enforcing the public water supply program in Iowa.

A "public water supply system" provides drinking water to 15 or more service connectors or to 25 or more individuals for more than 60 days per year.97/ Public water supply systems must obtain a DNR permit and must satisfy Maximum

94/ Iowa Admin. Code § 567-133.4(3)(b)(1).

95/ "An activity that does not violate chapter 455B does not violate this chapter [455E]." Iowa Code § 455E.6.

96/ 42 U.S.C. §§ 300f to 300j-26; 40 C.F.R. pts. 141-143.

97/ Iowa Code §§ 455B.177; Iowa Admin. Code § 567-40.2. Private water supplies generally are regulated by county boards of health.

Contaminant Levels (MCL) for pollutants in drinking water.98/ Iowa regulations impose obligations on public water supply systems to notify the DNR of all test results and to notify the public of violations of MCLs.99/ Like the groundwater protection requirements, MCLs are often important standards in Superfund cleanup proceedings described in chapter 5 of this handbook.

3.9 Well Closing Requirements

Iowa has adopted a number of specific regulations governing water wells.100/ In particular, an owner must plug a water well that is no longer in use or that is in such a state of disrepair that continued use for the purpose of accessing groundwater is unsafe or impracticable.101/ The DNR estimated in 1990 that there are approximately 70,000 abandoned wells to be plugged in Iowa.

DNR regulations specify the procedures for plugging a well and establish a priority schedule that phases in the well plugging requirements over 1990 to July 1, 2,000 depending on the type of well and other factors.102/ Violations may be subject to a civil penalty of $100 per every five calendar days that the well remains unplugged or improperly plugged, but not to exceed $1,000 and the penalty shall only be assessed after the $1,000 limit is reached.103/

98/ Iowa Code § 455B.183; Iowa Admin. Code § 567-40.2. MCLs are set in Iowa Admin. Code § 567-41.3 and are generally the same as the federal MCLs at 40 C.F.R. pt. 141. Notably, Iowa's MCLs for nitrates are more stringent.

99/ Iowa Admin. Code § 567-41.5; see also 40 C.F.R. § 141.32. The MCLs generally are expressed in milligrams per liter, or parts per million (mg/l = ppm). Measurements in micrograms per liter or parts per billion (ug/l = ppb) can be converted to ppm by moving the decimal point three digits. For example, 25,000 ppb equals 25 ppm.

100/ See Iowa Admin. Code chs. 567-38, 39, 47, 49.

101/ Iowa Code §§ 455B.171(24), 455B.190.

102/ Iowa Admin. Code ch. 567-39; see also Iowa Admin. Code §§ 567-110.12 (requiring closure of abandoned monitoring wells and boreholes near landfills), 567-135.8(q) (same at leaking underground storage tank sites).

103/ Iowa Code § 455B.190.

3.10 Other Areas of Water Regulation

In addition to the water pollution regulations discussed above, Iowa extensively controls other activities affecting its water resources, including: (1) water allocation, diversion, storage, withdrawal, and use,[104/] including priorities for farm over other uses;[105/] (2) flood plain control;[106/] (3) application of pesticides to waters;[107/] (4) standards and licensing requirements for commercial septic tank cleaners;[108/] (5) guidelines for private sewage disposal;[109/] and (6) regulation of levee and drainage districts.[110/]

3.11 Enforcement and Penalties

EPA and the DNR share enforcement authority over many water pollution laws in Iowa, including NPDES permits, although violations are enforced by the DNR in most situations. The DNR director may revoke or suspend NPDES permits for violation of any term or condition of such permits.[111/] In addition, violators of various water pollution statutes and regulations may be subject to both civil and

---

104/ See Iowa Code §§ 455B.261-282; Iowa Admin. Code chs. 567-50 to 54. As a general rule, a water permit is needed to withdraw 25,000 gallons or more per day of groundwater or surface water. Iowa Admin. Code § 567-51.6.

105/ "In the application for a permit to divert, store, or withdraw water and in the allocation of available water resources under a water permit system, the department of natural resources shall give priority to the use of water resources by a farm or farm operations, exclusive of irrigation, located in an agricultural area over all other uses except the competing uses of water for ordinary household purposes." Iowa Code § 352.11(2).

106/ See Iowa Code §§ 455B.261-455B.282; Iowa Admin. Code chs. 567-70 to 75.

107/ See Iowa Admin. Code ch. 567-66.

108/ See Iowa Admin. Code ch. 567-68.

109/ See Iowa Admin. Code ch. 567-69.

110/ Iowa Code ch. 468.

111/ Iowa Admin. Code § 567-64.3(11).

criminal penalties.112/ Under Iowa's citizen suit statute, any person adversely affected may commence a civil action against any person for violating Iowa's water protection statutes and rules.113/

3.11.1 Clean Water Act Penalties

Violations of NPDES permits may be enforced by the federal government in civil proceedings involving a potential $25,000 per day penalty in certain circumstances.114/ As an alternative to pursuing a civil penalty in federal court, after consultation with the DNR, the EPA may assess a class I or class II civil administrative penalty for NPDES violations.115/ A class I administrative penalty may not exceed $10,000 per violation or $25,000 total, and class II administrative penalty may not exceed $10,000 per day of violation or $125,000 total.116/

Negligent violations of the Clean Water Act, including NPDES permits, may be subject to a fine of not more than $25,000 per day of violation, one year imprisonment, or both.117/ Knowing violations may be punished by a fine of $50,000 per day of violation, three years imprisonment, or both. In addition, the EPA may impose criminal penalties under the Clean Water Act on "responsible corporate officers."118/ Finally, citizen suits may be brought for violations of effluent

112/ See Iowa Code §§ 455B.191, 455B.182. These statutes detail punishments for violations of Iowa Code §§ 455B.171-455B.192 (part 1 of division III of chapter 455B, which governs DNR jurisdiction) or for violations of rules promulgated under their authority.

113/ Iowa Code § 455B.111.

114/ 33 U.S.C. § 1319(a), (d).

115/ 33 U.S.C. § 1319(g)(1), (6)(A).

116/ 33 U.S.C. § 1319(g)(2).

117/ 33 U.S.C. § 1319(c).

118/ 33 U.S.C. § 1319(c)(6).

standards, including NPDES permits, if the EPA or the DNR are not already diligently prosecuting similar claims.[119]

3.11.2 Iowa Water Law Penalties

Violation of any provision of Iowa Code §§ 455B.171-455B.192 or rules promulgated under the authority of these statutes may be punished by a civil judicial penalty not to exceed $5,000 for each day of violation.[120] As discussed in section 1.2.3 of this handbook, the DNR may issue administrative penalties not to exceed $10,000 for violations of water quality statutes, rules, permits, or orders.[121] In recent years, the DNR has commonly pursued the administrative penalty method as an alternative to the civil judicial penalty method.

With regard to criminal enforcement, knowing or negligent commission of any of the following can be punished by both fine and imprisonment:

1. Violating Iowa Code § 455B.186, which prohibits certain discharges;

2. Violating Iowa Code § 455B.183, which governs requirements for written permits for discharge;

3. Violating conditions or limitations included in a permit issued under § 455B.183;

4. Introducing into a sewer system or publicly owned treatment works "any pollutant or hazardous substance which the person knew or reasonably should have known could cause personal injury or property damage;" or

5. Except as allowed by federal and state requirements and permits, causing a treatment works "to violate any water quality standard,

---

119/ 33 U.S.C. §§ 1319(g)(6), 1365; see also Arkansas Wildlife Fed'n v. ICI Americas, Inc., 1994 Westlaw 319,110 (8th Cir. 1994).

120/ Iowa Code § 455B.191(1).

121/ Iowa Code § 455B.109; Iowa Admin. Code § 567-10.3.

effluent standard, pretreatment standard or condition" of the treatment works' discharge permit.[122]

Negligent violators are guilty of a serious misdemeanor. For a first violation, negligent violators may be punished by a fine not to exceed $25,000 per day of violation, imprisonment for not more than one year, or both. Subsequent negligent violations by persons previously convicted of either negligent or knowing violations may be punished by fines of up to $50,000 per day of violation, imprisonment for not more than two years, or both.[123]

Knowing violators are guilty of an aggravated misdemeanor. For a first violation, knowing violators may be punished by a fine of not more than $50,000 per day of violation, by imprisonment for not more than two years, or by both fine and imprisonment. Subsequent knowing violations, by persons previously convicted of either negligent or knowing violations, may be punished by fines of not more than $100,000 per day of violatio, by imprisonment for not more than five years, or both.[124]

It is a crime to knowingly make misrepresentations on documents required by Iowa Code §§ 455B.171-455B.192 such as Discharge Monitoring Reports (DMR) or to knowingly render inaccurate any required monitoring equipment or method. Conviction may be punished by a fine of not more than $10,000, by imprisonment in the county jail for not more than six months, or both.[125]

Finally, it is prima facie evidence of contempt to fail to obey any order of the DNR related to a violation of Iowa Code §§ 455B.171-192 or rules or permits promulgated under their authority.[126] The DNR may certify to the district court of the county in which the contempt allegedly occurred. If the court finds that the

---

122/ Iowa Code § 455B.191(2).

123/ Iowa Code § 455B.191(2).

124/ Iowa Code § 455B.191(2).

125/ Iowa Code § 455B.191(3).

126/ Iowa Code § 455B.182.

DNR's order was lawful and reasonable, then it shall order compliance. If the person subject to the order fails to comply, the person is guilty of contempt and may be fined not more than five hundred dollars for each day of noncompliance with the court order.[127/]

127/ Iowa Code § 455B.182.

CHAPTER FOUR

WETLANDS AND PROTECTED WATERS

4.0 Introduction

Iowa's wetlands are protected by both federal and state regulatory programs. Wetlands, a subject of substantial national interest, are protected under the Clean Water Act and federal Swampbuster legislation. In addition, a regulatory scheme for wetlands was enacted by the Iowa Legislature in 1990.

4.1 Clean Water Act Section 404 Wetlands

The Clean Water Act attempts to control the discharge of "pollutants" into the nation's "navigable waters" by requiring a permit for these discharges. While the EPA has responsibility for issuing most Clean Water Act permits, wetlands are subject to the jurisdiction of the Army Corps of Engineers if they fall within the definition of "navigable waters of the United States" under Section 10 of the Rivers and Harbors Act of 1899 and implementing regulations,1/ or within the definition of "waters of the United States" under Section 404 of the federal Clean Water Act and implementing regulations.2/ Section 404 authorizes the Corps to issue permits "for the discharge of dredged and fill material into the navigable waters" of the United States.

"Waters of the United States" under the Clean Water Act include wetlands whose use, degradation, or destruction could affect interstate or foreign commerce as well as wetlands adjacent to other "waters of the United States."3/ The Clean Water Act has been broadly construed to extend the jurisdiction of the Corps to all water

---

1/ 33 C.F.R. pt. 329.

2/ 33 U.S.C. § 1344; 33 C.F.R. pt. 328.

3/ 33 C.F.R. § 328.3; 40 C.F.R. § 230.3(s); see United States v. Riverside Bayview Homes, Inc., 474 U.S. 121 (1985); Hoffman Homes, Inc. v. EPA, 999 F.3d 256 (7th Cir. 1993).

bodies, including wetlands, which can be shown to have a connection with interstate commerce, although this issue remains controversial. Even artificially created wetlands can be within the jurisdiction of the Corps.[4]

For purposes of the Section 404 permit, the term "wetlands" means those areas that are inundated or saturated by surface or groundwater at a frequency and duration sufficient to support, and that under normal circumstances do support, a prevalence of vegetation typically adapted for life in saturated soil conditions.[5] The criteria used to determine if an area of land meets this definition have been the subject of much debate in recent years.

Following the publication of United States Fish and Wildlife Service Circular No. 39 (1971 edition), which defines eight types of freshwater wetlands, the definition of "wetlands" has undergone considerable scrutiny. In 1987, the Corps adopted a Federal Manual for Identifying and Delineating Jurisdictional Wetlands (1987 manual). In November 1988, the National Wetlands Policy Forum, a group of governors, chief executives of environmental groups and corporations, real estate developers, ranchers, scientists, and federal officials, convened by the Conservation Foundation at the request of the EPA, issued its report with over 100 recommendations for wetlands protection. Among those recommendations was the adoption of a single regulatory definition of wetlands. The National Wetlands Policy Forum had identified 50 different definitions of a wetland in federal regulations. In January 1989, the EPA, the Corps, the Soil Conservation Service (SCS), and the Fish and Wildlife Service (FWS) signed an interagency agreement to adopt a single manual as the technical basis for identifying wetlands. This publication, the 1989 Federal Manual for Identifying and Delineating Jurisdictional

---

4/ Track 12, Inc. v. District Engineer, United States Army Corps of Engineers, 618 F. Supp. 448 (D. Minn. 1985).

5/ 33 C.F.R. § 328.3(b). Technically, a different definition of "wetlands" is included in Swampbuster which is discussed below.

Wetlands (1989 Manual), has been subject to substantial criticism on the grounds that it significantly increased the number of acres of lands considered to be wetlands and that it was adopted without notice and the opportunity for public comment.

In August 1991, the EPA, Corps, SCS, and FWS published notice of proposed revisions to the 1989 manual.6/ The public comment period was extended twice, and the EPA received over 70,000 comments on the proposed revisions.

The proposed revisions to the 1989 manual require an area to meet three criteria in order to be delineated as a wetland:

- Wetland hydrology;
- Hydric soil; and
- Hydrophytic vegetation.

Since some areas may meet only two of three wetland criteria, environmental groups have objected to the proposed revisions on the basis that as proposed, half of the remaining wetlands in the United States will not be protected.

The controversy surrounding the 1989 manual has not yet been resolved. An amendment to the 1992 Energy and Water Development Appropriations Act, signed by former President Bush, prohibits the Corps from applying the 1989 manual. As a result, the Corps now uses the original 1987 manual while EPA and other agencies rely on the 1989 manual. On August 24, 1993, the Clinton Administration unveiled its new federal wetlands program in anticipation of the Clean Water Act reauthorization process. That plan recommends the adoption of the 1987 Corps manual for use in wetlands delineation by all federal agencies.

There are no comprehensive maps identifying wetlands subject to the jurisdiction of the Corps. A site visit by the Corps is generally necessary to determine whether property contains a regulated wetland and to identify its dimensions.

6/ 56 Fed. Reg. 40446 (1991).

A second jurisdictional limitation on the authority of the Corps to regulate wetlands is created by the structure of the Clean Water Act. The Clean Water Act prohibits a "discharge" of a "pollutant" into the "waters of the United States."7/ "Pollutant" is defined to include dredge and fill material. Depositing fill material in a wetland is thus clearly within the Act. However, activities such as draining a wetland, removing the vegetation from a wetland, and dredging a wetland do not fall within a literal interpretation of "discharge." Several opinions have suggested that "discharge" should be broadly construed to achieve the purposes of the Clean Water Act and that such activities should be considered to be discharges if they cannot be accomplished without allowing some sediment to enter the water.8/

In August 1993, the Corps adopted the "excavation rule," which provides that "mechanized landclearing, ditching, channelization, and other excavation" activities constitute a discharge of dredged material.9/ This new rule also expands the authority of the Corps to regulate the placement of pilings in wetland areas.

4.2 Clean Water Act Section 404 Permit Program

As explained above, a permit from the Corps is required under Section 404 of the Clean Water Act to discharge dredge or fill material into waters of the United States within Iowa. Six activities are exempt from the need for a Section 404 permit, including certain farming, silviculture, and ranching operations, unless the discharge involves any toxic pollutant or the flow or circulation of waters of the United States may be impaired or the reach of such waters reduced.10/ Activities

---

7/ 33 U.S.C § 1311(e).

8/ See Weiszmann v. District Eng'r, U.S. Corps of Eng'rs, 526 F.2d 1302, 1305 (5th Cir. 1976).

9/ 33 C.F.R. § 323.2(d)(1).

10/ 33 C.F.R. § 323.4. The discharge of dredged or fill material is not prohibited "from normal farming, silviculture, and ranching activities such as plowing, seeding, cultivating, minor drainage, harvesting for the production of food, fiber, and forest products, or upland soil and

which are not exempt from the Section 404 permit program will generally require an individual, general, or nationwide permit.

An individual permit under Section 404 is a Corps authorization following a case–by–case evaluation of a specific project, including public notice and comment, and the opportunity for a public hearing.[11/] A general permit is a Corps authorization that is issued on a nationwide or regional basis for a category of activities, where the environmental consequences of the action are minimal and the activities are substantially similar in nature or the permit would avoid unnecessary duplication of regulation by other agencies.[12/] Nationwide permits are blanket authorizations for certain insignificant activities, such as placement of navigational aids and bank stabilization activities.[13/] Nationwide permit No. 26, for example, allows for certain discharges of dredged or fill material into "isolated waters."[14/]

When Congress enacted the Section 404 program in 1972, it gave permitting authority to the Corps but recognized the EPA's responsibility for environmental protection and thus gave the EPA authority to develop guidelines for the specification of disposal sites or the issuance of permits, known as the 404(b)(1) guidelines.[15/] The Corps must follow these guidelines in determining whether to grant a permit under Section 404. Under Section 404(c), the EPA also has veto authority over the Corps' grant of any permit where it finds that the discharge

---

water conservation projects." 33 U.S.C. § 1344(f)(1)(a). Courts generally have interpreted this provision narrowly. See United States v. Huebner, 752 F.2d 1235 (7th Cir. 1985); United States v. Cumberland Farms, 826 F.2d 1151 (1st Cir. 1987).

11/ 33 C.F.R. pts. 325, 327.

12/ 33 C.F.R. § 323.3(h).

13/ 33 C.F.R. § 330.5.

14/ 33 C.F.R. Part 330, App. A.

15/ 40 C.F.R. pt. 230.

would result in unacceptable adverse effects. States also are involved in the program through Section 401 which requires the permit applicant to provide the state's certification that the discharge will comply with the Clean Water Act.

The decision to issue a permit will be based on an evaluation of the probable impacts of the proposed activity, and "[t]he benefit which reasonably may be expected to accrue from the proposal must be balanced against its reasonably foreseeable detriments."16/ In this balancing of benefits and detriments, mitigation of adverse impacts is often necessary for a permit to be issued by the Corps. Acceptable mitigation generally turns on replacing the value of the wetland. The mitigation policy of the United States Fish and Wildlife Service, which functions in an advisory role in the Section 404 program, also focuses on habitat value.17/ An applicant for a permit, however, is first expected to design the project to avoid and minimize encroachment on the wetland, review alternatives to wetland alteration, and assess the extent of impacts which cannot be avoided.

4.3 Clean Water Act Section 404 Enforcement

4.3.1 Civil Enforcement

The Corps and the EPA share enforcement authority for the Clean Water Act Section 404 program. The Corps and the EPA have interagency agreements with respect to this shared authority. Following an initial investigation, the Corps may issue a cease and desist order involving violations of Section 404 where a project is not yet complete. Where the project has been completed, the Corps may issue an order for initial corrective measures taking into account jeopardy to life, property, or important public resources. Except where the Corps has determined that legal action is appropriate, it will review an after–the–fact permit application which will be

16/ 33 C.F.R. § 325.3(c)(1).

17/ 46 Fed. Reg. 7644 (Jan. 23, 1981).

granted if the work is not contrary to the public interest and complies with the EPA's Section 404(b)(1) guidelines. If an after–the–fact permit is denied, notification of the denial may be accompanied by the required final corrective actions for necessary restoration work. Failure to complete the restoration likely will result in a legal action.18/

Civil penalties for Section 404 violations may not exceed $25,000 per day of violation. In 1987, Congress authorized issuance of Class I (not to exceed $25,000) and Class II (not to exceed $125,000) administrative penalties for violations of Section 404.19/ Structures, fill, deposit or other flood use in the floodplain in violation of a floodplain management ordinance is a public nuisance which may be enjoined by a judicial action brought by the DNR or a local unit of government.20/

4.3.2 Criminal Enforcement

The Corps also may recommend a criminal action for violation of the Section 404 permit requirements. Knowing violations of the Section 404 requirements became felonies following amendments to Section 309 of the Clean Water Act in 1987. Federal criminal penalties for a knowing violation may not exceed $50,000 per day of violation and/or 3 years imprisonment.21/

4.4 Federal Swampbuster Wetlands Provisions

The Food Security Act of 1985 was the first federal law to create severe restrictions for farmers on draining or otherwise altering wetlands.22/ In 1990, the

---

18/ 33 C.F.R. pt. 326.

19/ 33 U.S.C. § 1319(g).

20/ For further discussion of nuisance actions, see chapter 14 of this handbook.

21/ 33 U.S.C. § 1319(c).

22/ See 16 U.S.C. §§ 3801–3845.

U.S. Department of Agriculture (USDA) published final regulations implementing the Swampbuster provisions.23/ Swampbuster prohibits farmers who produce commodities on a wetland converted after December 23, 1985, or who converted wetlands into cropland after November 28, 1990, from participating in most USDA programs.24/ Although Swampbuster does not make it illegal to convert wetlands to agricultural use, it establishes powerful financial incentives to comply.

4.4.1 Wetland Determination Process Under Swampbuster

The determination of whether an area is a "wetland" for purposes of the Swampbuster law is a technical decision made by the district conservationist of the USDA Soil Conservation Service. While the Swampbuster definition of the term "wetland" is contained in SCS regulations, the adoption of the 1989 Manual by EPA, Corps, SCS and other federal agencies altered the application of the regulatory definition. For this reason, the SCS in December, 1991 proposed revising the Swampbuster regulatory definition of the term "wetland" to correspond to proposed revisions in the 1989 Manual.

The first step in the SCS wetland determination process is for the district conservationist to review aerial photographs and wetland maps to identify the location of wetlands.25/ The final determination of whether an area meets wetland criteria can be made based upon existing records and without an on–site determination. However, on–site determinations may be requested and site visits must be conducted if the owner has disagreed with the determination or if adequate information is not available.

23/ 7 C.F.R. § 12.

24/ 16 U.S.C. § 3821; 7 C.F.R. § 12.4(a). Conversion of wetlands, with some exceptions, results in a farmer's loss of eligibility for eleven federal programs, including among others price supports, disaster payments, crop insurance, and FmHA loans. 7 C.F.R. § 12.4(c).

25/ 7 C.F.R. § 12.6(c)(4).

4.4.2 Appeal of SCS Wetland Determinations

According to Swampbuster regulations: "Persons who are adversely affected by a determination made under this section [SCS procedures] and believe that the requirements of this part were improperly applied may appeal . . . any determination by SCS." Appeals of wetland determinations are made to SCS under its regular administrative appeals procedure.26/ The SCS appeals process begins with reconsideration by district SCS office, with subsequent appeals to the area conservationist, the state conservationist, and the SCS chief in Washington, D.C.27/

4.4.3 Exempt Wetlands Under Swampbuster

Some wetlands are exempt from Swampbuster regulation. Most importantly, Swampbuster provisions do not apply to wetlands on which a conversion to cropland was completed or commenced before December 23, 1985.28/ The term "converted wetland" is defined in SCS regulations as:

> wetland that has been drained, dredged, filled, leveled, or otherwise manipulated (including any activity that results in impairing or reducing the flow, circulation, or reach of water) that makes possible the production of an agricultural commodity without further application of manipulations described herein if (i) such production would not have been possible but for such an action; and (ii) before such an action land was wetland and was neither highly erodible land nor highly erodible cropland.29/

4.4.4 "Farmed Wetlands" Under Swampbuster

Land currently used to grow commodities and which was converted prior to December 23, 1985 may still be considered a "farmed wetland" under Swampbuster if the land meets the soil and water criteria and would support the growth of wetlands

---

26/ 7 C.F.R. §§ 12.6(c)(5), 12.12.

27/ 7 C.F.R. § 614.

28/ 7 C.F.R. § 12.5(b)(i).

29/ 7 C.F.R. § 12.2(a)(6).

vegetation if farming was abandoned. A "farmed wetland" may only be farmed as it was prior to December 23, 1985 and no new drainage which would bring additional wetland into production may be constructed on it.30/ Farmers may maintain drainage systems on farmed wetland but cannot expand the "scope and effect" of the original drainage system without becoming ineligible for program benefits.31/

4.4.5 ASCS Oversight of Wetlands Under Swampbuster

The Agricultural Stabilization and Conservation Service is charged with the implementation and enforcement of Swampbuster.32/ ASCS county committees consider USDA program eligibility and loss of benefits.33/ The loss of farm program benefits is appealed under ASCS appeal procedures which are similar to the SCS appeal procedures.

4.5 Iowa Wetlands Regulation

In 1990, the State of Iowa enacted legislation requiring the Iowa DNR to regulate the drainage of certain "protected wetlands" and to undertake a county–by–county inventory of wetlands in the state.34/ The definition of "wetland" adopted in the 1990 wetlands legislation is limited to "an area of two or more acres in a natural condition that is mostly under water or waterlogged during the spring growing season and is characterized by vegetation of hydric soils."35/ "Protected wetlands" includes United States Department of Interior Circular 39 type

---

30/ See 7 C.F.R. § 12.33(b).

31/ 7 C.F.R. § 12.33(b).

32/ 7 C.F.R. § 12.6(b).

33/ 7 C.F.R. § 12.6(b).

34/ See generally Note, Federal and State Regulation of Wetlands in Iowa, 41 Drake L. Rev. 139 (1992).

35/ Iowa Code § 456B.1(5).

3, type 4, and type 5 wetlands.[36] Under the Iowa statute, "protected wetlands" do not include "land where an agricultural drainage well has been plugged causing a temporary wetland" or "land within a drainage district or levee district."[37] This exemption along with the two-acre minimum size criterion provides a significant exemption from coverage for areas that otherwise may be considered wetlands.

4.5.1 Iowa Protected Wetlands Inventory

The Iowa statute provides that the DNR "shall inventory the wetlands and marshes of each county and make a preliminary designation as to which constitute 'protected wetlands.'"[38] This designation process takes place in three steps.

First, preliminary "protected wetland" determinations will be made by the DNR in consultation with county conservation boards. The DNR is then required to prepare county maps showing areas to be designated as "protected wetland." The map is to be filed with the county conservation board and county recorder.[39]

Second, the DNR is to notify affected landowners of the preliminary wetland designations by certified mail. The notice must contain information on how the landowner can challenge the designations of certain areas as "protected wetlands" or "request the designation of additional marshes or wetlands as protected wetlands."[40] The affected landowner may contest the preliminary designation in a hearing with the DNR director or by participating in mediation conducted by the Iowa Farm Mediation Service.[41]

36/ Iowa Code § 456B.1(4).

37/ Iowa Code § 456B.1(4).

38/ Iowa Code § 456B.12.

39/ Iowa Code § 456B.12.

40/ Iowa Code § 456B.12.

41/ Iowa Code §§ 456B.12, 654A.16.

Finally, following the hearing or mediation, the director of the DNR will issue an order designating the protected wetlands in the county.42/ This order is considered a final order of the DNR and can be appealed to the district court pursuant to the Iowa Administrative Procedure Act.

4.5.2 Wetland Drainage Permit

The Iowa act prohibits the draining of a "protected wetland" without a permit issued by the DNR.43/ The statute provides that the DNR shall not issue a permit to drain a protected wetland except under one of the following conditions:

1. A protected wetland is replaced by the applicant with a wetland of equal or greater value as determined by the DNR; or

2. The protected wetland does not meet the criteria for continued designation as a protected wetland.44/

Notwithstanding these limitations, however, the Act provides that a landowner may utilize "the bed of a protected wetland for pasture or cropland if there is no construction of dikes, ditches, tile lines or buildings and the agricultural use does not result in drainage."45/

4.5.3 Penalties

The Iowa act creates a civil penalty for violating the permit requirements.46/ Fines of not more that five hundred dollar for each day that the violation continues

---

42/ Iowa Code § 456B.12.

43/ Iowa Code § 456B.13.

44/ Iowa Code § 456B.13.

45/ Iowa Code § 456B.13.

46/ Iowa Code § 456B.14.

may be assessed. A civil penalty does not take effect until the fourth day after the landowner is given written notification of the violation.47/

4.6 Iowa's Nonregulatory Protection of Wetlands

4.6.1 Tax Exemption of Wetlands

Land designated as "protected wetland" by the DNR and wetlands which were previously drained and farmed and then subsequently restored pursuant to a nonpermanent restoration agreement with the regulatory authority may qualify for a property tax exemption in Iowa.48/ An application for the exemption may be made by the landowner on forms provided by the Iowa Department of Revenue. The application forms must be filed with the applicable assessing authority by the first of February in the year the exemption is requested. If the applicable land qualifies, the DNR will issue a exemption from property tax assessment. Property receiving the exemption may not be used for "economic gain" during the exemption period.49/

4.6.2 Highway Construction in Wetlands

Iowa's Department of Transportation is required to mitigate the destruction of wetlands. Iowa Code § 314.23 provides:

> It is declared to be in the general public welfare of Iowa and a highway purpose that highway maintenance, construction, reconstruction, and repair shall protect and preserve, by not causing unnecessary destruction, the natural or historic heritage of the state. In order to provide for the protection and preservation, the following shall be accomplished in the design, construction, reconstruction, relocation, repair, or maintenance of roads, streets, and highways:

* * *

47/ Iowa Code § 456B.14.

48/ Iowa Code § 427.37.

49/ Iowa Code § 427.37.

2. Wetlands. Wetland removed shall be replaced by acquisition of wetland, in the same general vicinity if possible, for public ownership and preservation, or by other mitigation deemed to be comparable to the wetland removed, including, but not limited to, the improvement, development, or preservation of wetland under public ownership.50/

4.7 The "Takings" Question

The Fifth Amendment to the United States Constitution and Article I, Section 18 of the Iowa Constitution prohibit the taking of private property for a public use without paying just compensation. With the expanding regulation of wetlands and other protected areas, the question of when regulation becomes an unconstitutional taking has taken on greater interest.

The U.S. Supreme Court has held that regulations can cause a taking of property.51/ In the wetlands context, several claims involving regulatory takings by the Corps have been brought in the U.S. Court of Federal Claims.52/ In a recent case, the U.S. Supreme Court reviewed a South Carolina law that required owners of "critical areas" coastal zone land to obtain a permit before changing the use of the land and prohibited construction of habitable improvements seaward of a new baseline.53/ The Court held that for a state to avoid a "takings" claim where a regulation deprives land of all beneficial use, the state must demonstrate that the owner's use of the property was similarly restricted by common law principles, separate and apart from a newly adopted regulation.

50/ Iowa Code § 314.23.

51/ Nollan v. California Coastal Comm'n, 483 U.S. 825, 837 (1987); First Evangelical Lutheran Church v. County of Los Angeles, 482 U.S. 304, 319 (1987).

52/ Loveladies Harbor Inc. v. United States, 28 F.3d 1171, 1182-83 (Fed. Cir. 1994) (affirming $2,658,000 judgment based on regulatory taking of 12.5 acres of wetlands); Florida Rock Indus., Inc. v. United States, 18 F.3d 1560, 1567 (Fed. Cir. 1994) (vacating $1,029,000 judgment and remanding); Deltona Corp. v. United States, 657 F.2d 1184 (Cl. Ct. 1981) (developer did not suffer uncompensated taking), cert. denied, 455 U. S. 1017 (1982).

53/ Lucas v. South Carolina Coastal Council, 112 S. Ct. 2886 (1992).

More recently, the U.S. Supreme Court reviewed a conditional use permit that required a landowner to dedicate a portion of land within a floodplain for storm drainage and land adjacent to the floodplain for a bicycle path as a condition of expansion of a commercial building.[54] The Court ruled that these conditions had a legitimate public purpose, but remanded for further information regarding whether the conditions had a "reasonable relationship" or "rough proportionality" to the projected impact of the development. If not, the conditions amount to a compensable "taking." Similarly, under the Iowa Constitution the Iowa Supreme Court has ruled that regulation effectively prohibiting all economically beneficial use of land results in a compensable "taking."[55]

4.8 Drainage Wells

In 1990, Iowa enacted the Groundwater Protection Act to address the problems associated with agricultural chemicals entering the groundwater through agricultural surface drainage.[56] The law provides that the DNR is to take efforts to eliminate "chemical contamination caused by the use of agricultural drainage wells by January 1, 1995."[57] Under the law, the State is to identify and educate farmers on alternative farm management practices to reduce the "infiltration of synthetic organic compounds into the water through agricultural drainage wells and sinkholes."[58] The DNR is also required to implement a program "for the

---

54/ Dolan v. City of Tigard, 114 S. Ct. 2309, 2316-17 (1994).

55/ Iowa Coal Min. Co. v. Monroe County, 494 N.W.2d 664 (Iowa 1993), cert. denied, 113 S. Ct. 2415 (1993).

56/ Iowa Code § 159.29.

57/ Iowa Code § 159.29(7).

58/ Iowa Code § 159.29(3)(b).

acquisition of wetlands and conservation easements on or around wetlands that result from the closure or change in use of agricultural drainage wells."[59]

The Act also requires owners of agriculture drainage wells to register the well with the DNR.[60] In addition, an owner of a drainage well must work with the DNR to develop drainage alternatives.[61] Landowners who do not register agricultural drainage wells or do not develop alternative draining plans are ineligible to receive state money in the agricultural management account of the groundwater protection fund.[62]

4.9 Protected Water Areas

In 1984, Iowa enacted the Protected Water Area System Act to help protect and preserve designated Iowa "water areas."[63] Water areas which may be protected under the Act must:

> possess outstanding cultural and natural resource values such as water conservation, scenic, fish, wetland, forest, prairie, mineral, geological, historic, archaeological, recreation, education, water quality or flood protection values.[64]

If these characteristics exist, the Natural Resources Commission, other public agencies, interest groups and private parties can nominate a water area for

---

59/ Iowa Code § 456B.11.

60/ Iowa Code § 159.29(1).

61/ Iowa Code § 159.29(2).

62/ Iowa Code § 159.29(2)(b).

63/ Iowa Code ch. 462B. "Water area" means "a river, lake, wetland, or other body of water and adjacent lands where the use of those lands affects the integrity of the water resource." Iowa Code § 462B.1(10).

64/ See Iowa Code § 462B.1(6).

classification as a "protected water area system."65/ After the nomination of prospective protected water areas, but prior to the designation of the water area as a "prospective protected water," the Commission must conduct a public hearing.66/ Seven days prior to the hearing, the Commission must provide notice in a newspaper having general circulation in the county where the proposed water area is located.67/ The Commission is not required to give any personal notice to landowners and others that may be directly affected by the protected water designation.68/ Following the public hearings, the Commission may designate all or part of the water area as a "prospective protected water area."69/

The "prospective" designation may remain in effect for up to two years while the Commission develops a management plan for the "protection and enhancement" of the water area.70/ The management plan contains specific recommendations for the development, management, and use of the protected water area. The Commission is authorized to use any combination of existing means, except condemnation, for managing and preserving the water area.71/

Iowa Code section 462B.10 encourage cooperation in this process:

> Recognizing that most of the protected water areas may be within privately owned lands, the legislature encourages the commission to cooperate with

---

65/ See Iowa Code § 462B.3. "Protected Water Area System" is defined as "a total comprehensive program that includes the goals and objectives, the state plan, the individual management plans, the prospective protected water areas, the protected water areas, the acquisition of fee title and conservation easements and other agreements, and the administration and management of such areas." Iowa Code § 462B.1(8).

66/ Iowa Code § 462B.5.

67/ Iowa Code § 462B.5.

68/ Upper Iowa River Preservation Ass'n v. Iowa Natural Resource Comm'n, 497 N.W2d 865, 868-69 (Iowa 1993).

69/ Iowa Code § 462B.4.

70/ Iowa Code § 462B.4.

71/ Iowa Code § 462B.9

> the landowners within the designated in achieving the purposes of this chapter. Likewise, the landowners within the designated areas are encouraged to cooperate with the commission. Commission staff shall meet separately or in small groups with landowners within interim protected water areas during the preparation of the master plan to establish workable and acceptable agreements for the protection of the area and its accompanying resources in a manner consistent with the purposes of this chapter and the interest and concerns of the landowner.

The apparent purpose of this provision is to promote consensus and voluntary agreements between the Commission and persons affected by the protected water designation regarding both the areas to be designated for protection and the management plan for the area. In Upper Iowa River Preservation Ass'n v. Iowa Natural Resources Commission,[72/] the Iowa Supreme Court held that this provision is directory rather than mandatory and therefore it is not necessary for the Commission to meet individually or in small groups with affected persons or enter into voluntary protection agreements with landowners before a management plan is adopted.[73/] Apparently the Commission can satisfy the statutory requirements by simply publishing notice of the management plan "at least twice" in a newspaper of general circulation and holding a final public hearing on the completed management plan.[74/]

At least thirty days following the public hearings, the Commission may permanently designate the water area a "protected water area" and adopt a management plan which may be submitted to the Iowa legislature for funding consideration.[75/] In addition, Iowa's Resource Enhancement and Protection Act

---

72/ 497 N.W.2d 865 (Iowa 1993).

73/ 497 N.W.2d at 869–70. The Court also ruled that the Commission could enter into voluntary protection agreements after the management plan was adopted.

74/ Iowa Code § 462B.7.

75/ Iowa Code § 462B.8.

(REAP) allocates funding to protected water areas.76/ The Commission may request that this REAP money be used to develop, manage, or preserve a protected water area system.

4.10 Calcareous Fens

A calcareous fen is a peat-accumulating wetland dominated by neutral to alkaline waters with high concentrations of calcium and low dissolved oxygen. Fens provide an environment for certain rare types of plants. Fen protection has been very controversial in the midwest, and some states have adopted special legislation to protect fens.77/ In Iowa, fens are protected by heightened enforcement of the general wetlands laws. According to a statewide inventory, 200 fens are located in the 37 Iowa counties indicated in the following figure.

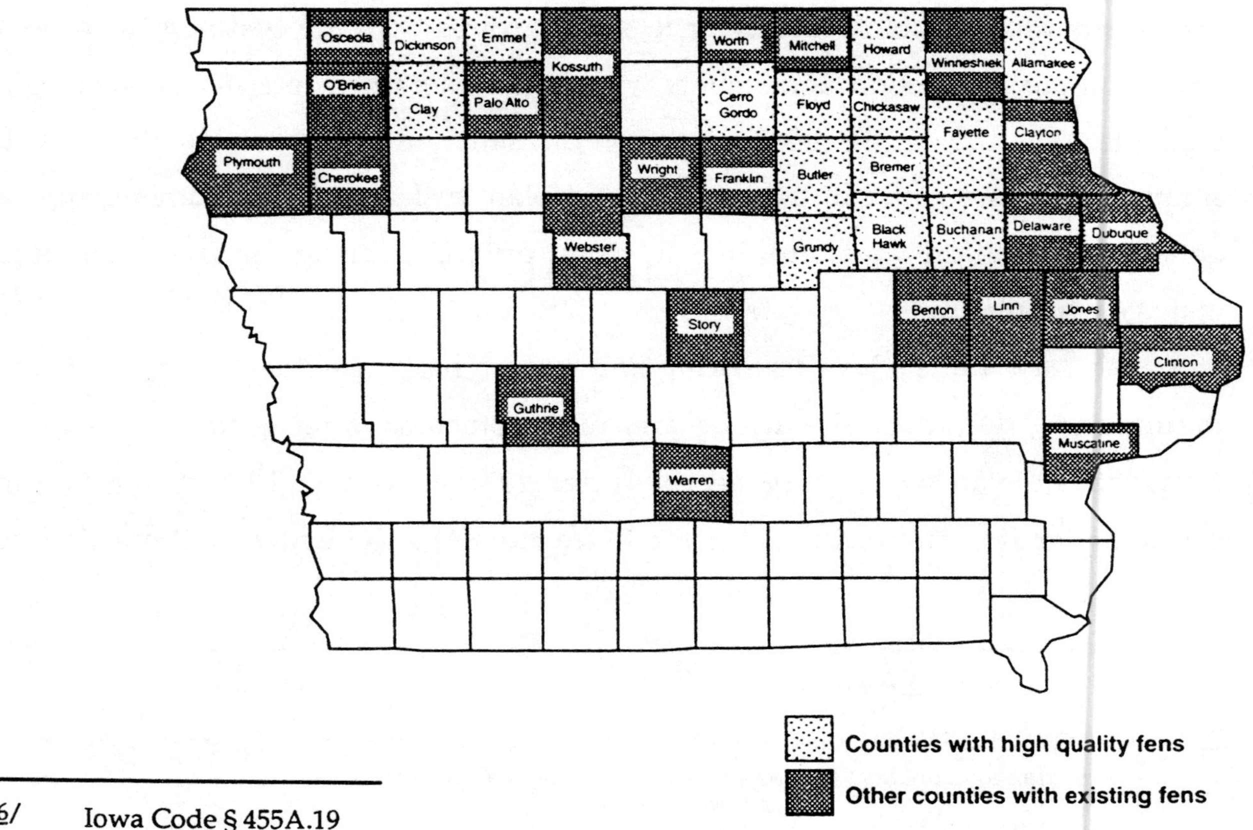

76/ Iowa Code § 455A.19

77/ Minn. Stat. § 103G.223; Minn. R. pts. 8420.1010-8420.1060.

CHAPTER FIVE

SUPERFUND: HAZARDOUS CONDITIONS CLEANUP

5.0 General Features of Superfund Laws

The federal Superfund law, formally known as the Comprehensive Environmental Response, Compensation and Liability Act of 1980 as amended by the Superfund Amendments and Reauthorization Act (SARA) of 1986 (together known as CERCLA),1/ was enacted to identify and clean up abandoned and existing sites contaminated by hazardous substances. The Superfund law encourages persons responsible for a release to undertake investigations and cleanups of the contaminated sites although the government is authorized to take action in response to the release as well. Thus, if a responsible party fails to undertake remedial actions or the party cannot be found, the government can undertake the work and seek reimbursement of its costs from responsible parties.

Iowa has adopted a counterpart to the federal Superfund law that is more limited in some ways, but broader in other respects. The Iowa statutes provide a mechanism for cleanup of hazardous conditions, which generally include sites contaminated with hazardous substances. Because much of the case law involving contaminated sites arises under the federal Superfund law and would likely apply under the Iowa statutes as well, CERCLA and Iowa law are discussed together in this chapter.

The Superfund law's popular name is derived from the fund it established to facilitate the investigation and cleanup of contaminated sites.2/ The federal Superfund is financed with special taxes on crude oil, petroleum products, chemical feedstocks, and corporate profits as well as with money recovered in cost recovery actions. To finance the 10% state share of abandoned site cleanup costs and other cleanup activities, the Iowa legislature established the Hazardous Waste Remedial

---

1/ 42 U.S.C. §§ 9601-9675.

2/ 26 U.S.C.§ 9507; 42 U.S.C. § 9601(11).

Fund.3/ Both federal and Iowa law, however, are based on the principle that the "polluter pays" and Superfund funding of cleanups is rare.

Prominent features of the federal Superfund law include the following:

1. It imposes joint and several liability on responsible persons;4/

2. It imposes strict liability, that is, liability without regard to fault or negligence;5/

3. It imposes liability retroactively; that is, persons responsible for contamination prior to the enactment of the Superfund law are nonetheless liable;6/ and

4. It imposes severe penalties for noncompliance with statutory requirements or government orders in appropriate cases.7/

3/ Iowa Code § 455B.423.

4/ See In re Bell Petroleum Servs., Inc., 3 F.3d 889, 902-04 (5th Cir. 1993); United States v. Aceto Agri. Chem. Corp., 872 F.2d 1373, 1377 (8th Cir. 1989); United States v. Northeastern Pharmaceutical & Chem. Co. (NEPACCO), 579 F. Supp. 823, 844-45 (W.D. Mo. 1984), aff'd, 810 F.2d 726 (8th Cir. 1986), cert. denied, 484 U.S. 848 (1987). Joint and several liability means that each of the parties responsible for a release is responsible for the entire cost of the investigation and cleanup. Thus, a single small generator of hazardous substances deposited on a contaminated site could be responsible for all remedial actions if other larger generators are bankrupt or out of business.

5/ 42 U.S.C. § 9601(32); NEPACCO, 810 F.2d at 743; United States v. Marisol, Inc., 725 F. Supp. 833, 839 (M.D. Pa. 1989).

6/ United States v. Reilly Tar & Chem. Corp., 546 F. Supp. 1100, 1113 (D. Minn. 1982); New York v. Shore Realty Corp., 759 F.2d 1032, 1043 (2d Cir. 1985); United States v. Monsanto Co., 858 F.2d 160, 173-75 (4th Cir. 1988), cert. denied, 109 S. Ct. 3156 (1989).

7/ See, e.g., 42 U.S.C. § 9607(c)(3); Solid State Circuits, Inc. v. U.S. EPA, 812 F.2d 383, 390-92 (8th Cir. 1987); United States v. Carolina Transformer Co., 739 F. Supp. 1030, 1039 (E.D. N.C. 1989), aff'd, 978 F.2d 832 (4th Cir. 1992); United States v. Parsons, 723 F. Supp. 757, 763-65 (N.D. Ga. 1989), vacated on other grounds, 936 F.2d 526 (11th Cir. 1991).

Subject to limited defenses, the general rule of liability under Superfund is that a person is responsible for a release of a hazardous substance from a facility[8/] if the person:

1. Owned or operated the facility when it was contaminated;

2. Owned or possessed a hazardous substance, or pollutant or contaminant, and arranged for its disposal, treatment, or transportation; or

3. Knowingly accepted a hazardous substance, or pollutant or contaminant, and (a) transported it to a facility selected by the transporter or (b) disposed of it in a matter contrary to law.[9/]

These three categories of responsible persons are generally referred to as owners or operators, generators or arrangers, and transporters, respectively. These concepts generally apply to the Iowa hazardous conditions statutes although the Iowa statutes do not expressly provide for private causes of action. As a result, many disputes between private parties relating to contaminated sites are litigated under the federal Superfund law in Iowa.

5.1 Releases, Hazardous Substances, and Hazardous Conditions

The concept of a "release" of a "hazardous substance" is central to both the federal and Iowa Superfund schemes because the responsible persons are liable only in the event of a release or threatened release of a hazardous substance. Under the Iowa statutes, "release" means "a threatened or real emission, discharge, spillage, leakage, pumping, pouring, emptying, or dumping of a hazardous substance into or

---

8/ The term "facility" means "(A) any building, structure, installation, equipment, pipe or pipeline (including any pipe into a sewer or publicly owned treatment works), well, pit, pond, lagoon, impoundment, ditch, landfill, storage container, motor vehicle, rolling stock, or aircraft, or (B) any site or area where a hazardous substance has been deposited, stored, disposed of, or placed, or otherwise come to be located; but does not include any consumer product in consumer use or any vessel." 42 U.S.C. § 9601(9).

9/ See 42 U.S.C. § 9607(a).

onto the land, air, or waters of the state" unless certain exceptions apply.10/ The CERCLA definition at 42 U.S.C. § 9601(20) is similar.

The Iowa statutes at Iowa Code § 455B.381 define "hazardous substance" as follows:

> Any substance or mixture of substances that presents a danger to the public health or safety and includes, but is not limited to, a substance that is toxic, corrosive, or flammable, or that is an irritant or that generates pressure through decomposition, heat, or other means. "Hazardous substance" may include any hazardous waste identified or listed by the administrator of the United States environmental protection agency under the Solid Waste Disposal Act as amended by the Resource Conservation and Recovery act of 1976, or any toxic pollutant listed under section 307 of the federal Water Pollution Control Act as amended to January 1, 1977, or any hazardous substance designated under section 311 of the federal Water Pollution Control Act as amended to January 1, 1977, or any hazardous material designated by the secretary of transportation under the Hazardous Material Transportation Act.

In contrast to the CERCLA definition of "hazardous substance," the Iowa definition does not expressly exclude petroleum.11/ Under these broad definitions, regulatory

---

10/ Iowa Code § 455B.381 (9). The exceptions apply when the release is in compliance with a federal or state permit, in accordance with the product label for the use of a pesticide, or the "hazardous substance is confined and expected to stay confined to property owned, leased or otherwise controlled by the person having control over the hazardous substance." Id.

11/ "Hazardous Substance" is defined in 42 U.S.C. § 9601(14) to mean:

1. Any substance designated pursuant to section 1321(b)(2)(A) of the federal Clean Water Act, 33 U.S.C. §§ 1251-1387 (These substances are listed at 40 C.F.R. § 116.4.);

2. Any element, compound, mixture, solution, or substance designated pursuant to section 9602 of CERCLA (This list is at 40 C.F.R. § 302.4.);

3. Any hazardous waste having the characteristics identified under or listed pursuant to section 3001 of the Resource Conservation and Recovery Act (RCRA), 42 U.S.C. §§ 6901-6992k; (These lists are at 40 U.S.C. § 261.30. The characteristics are ignitability (defined at 40 C.F.R. § 261.21), corrosivity (40 C.F.R. § 261.22), reactivity (40 C.F.R. § 261.23) and toxicity (40 C.F.R. § 261.24). "Hazardous substance" under CERCLA does not include any waste the regulation of which under RCRA has been suspended by Congress);

4. Any toxic pollutant listed under section 1317(a) of the Clean Water Act;

agencies take the view that once the hazardous substance is discharged, the release continues until the substance is abated. Thus, any person falling into the category of "responsible person" within the meaning of CERCLA over the entire time the substance remains in the environment may be deemed liable under CERCLA for its cleanup.

Another key concept under the Iowa statutes is the term "hazardous condition." This term is broadly defined in Iowa Code § 455B.381(4) as:

> [A]ny situation involving the actual, imminent, or probable spillage, leakage, or release of a hazardous substance onto the land, into a water of the state, or into the atmosphere, which creates an immediate or potential danger to the public health or safety or to the environment. For purposes of this division, a site which is a hazardous waste or hazardous substance disposal site as defined in section 455B.411, subs. 4, is a hazardous condition.

Liability under the Iowa statute is triggered by the presence of a hazardous condition.12/ Similarly, the obligation to notify the DNR discussed in chapter 8 of this handbook depends on occurrence of a hazardous condition.13/

5.2 Liability of "Responsible" Persons

5.2.1 Liability of Owners or Operators, Arrangers, and Transporters

As noted above, responsible parties under CERCLA include owners,

---

5. Any hazardous air pollutant listed under section 741 the Clean Air Act, 42 U.S.C. §§ 7401-7642; and

6. Any imminently hazardous chemical substance or mixture with respect to which the Administrator has taken action pursuant to section 2606 of the Toxic Substance Control Act, 15 U.S.C. §§ 2601-2671.

The definition does not include natural or synthetic gas or petroleum, crude oil or any fraction thereof. 42 U.S.C. § 9601(14); see also Wilshire Westwood Assocs. v. Atlantic Richfield Co., 881 F.2d 801, 804-05 (9th Cir. 1989) (leaded gasoline is not a hazardous substance). Moreover, it does not include vehicle emissions or proper applications of agricultural chemicals.

12/ Iowa Code § 455B.392.

13/ Iowa Code § 455B.386.

operaters, arrangers,[14/] and transporters. Responsible persons are frequently referred to as "potentially responsible parties" or PRPs.

A prominent Iowa case involving the Aceto Agricultural Chemicals Corp. site in Mills County led to an Eighth Circuit decision interpreting these CERCLA categories broadly. In the Aceto case, pesticide manufacturers contracted with a formulater to formulate pesticides while retaining ownership of the pesticides and with knowledge that spills would result during the formulation process. Under these circumstances, the Eighth Circuit upheld the manufacturers' liability as "arrangers" under CERCLA.[15/]

Subject to certain exclusions and defenses, a responsible person under Iowa law is a "person having control over a hazardous substance." This term is defined in Iowa Code § 455B.381(8) as:

> [A] person who at any time produces, handles, stores, uses, transports, refines, or disposes of a hazardous substance the release of which creates a hazardous condition, including bailees, carriers, and any other person in control of a hazardous substance when a hazardous condition occurs, whether the person owns the hazardous substance or is operating under a lease, contract, or other agreement with the legal owner of the hazardous substance.

5.2.2 Liability of Employees, State Employees, and "Good Samaritans"

Under Iowa law, it is unclear whether an employee would be regarded as a person having control over a hazardous substance. The general rule in most jurisdictions is that an employee is subject to liability only if the employee's conduct with regard to the hazardous substance was negligent under circumstances in which the employee knew that the substance was hazardous and that that conduct, if

---

14/ "Arrangers" are persons who arrange for the disposal, treatment, or transportation of a hazardous substance. Courts have generally required some affirmative action by the defendant for "arranger" liability to attach. Ashland Oil, Inc. v. Sonford Products Corp., 810 F. Supp. 1057, 1061 (D. Minn. 1993); United States v. Consolidated Rail Corp., 729 F. Supp. 1461, 1469-70 (D. Del. 1990) see also United States v. Arrowhead Refining Co., 829 F. Supp. 1078, 1090 (D. Minn. 1992) (oil company not liable for used oil disposal by service station lessee).

15/ United States v. Aceto Agri. Chem. Corp., 872 F.2d 1373, 1382 (8th Cir. 1989).

negligent, could result in harm. Employers will likely be held responsible regardless of the degree of care exercised by the employee.

Iowa statutes provide that state employees are not liable for damages incurred as a result of actions taken when acting in the person's official capacity pursuant to the Iowa hazardous conditions statutes.16/ A "good Samaritan" who provides assistance at the request of the DNR is not liable in a civil action for damages as a result of that person's acts or omissions provided that the person is not otherwise a responsible party, did not obtain payment for its services, and did not commit intentional wrongdoing or gross negligence.17/ Similarly, a person who provides assistance or advice in mitigating or attempting to mitigate the effects of an actual or threatened hazardous condition or in preventing, cleaning up, or disposing -- or in attempting to prevent, clean up, or dispose -- of a hazardous condition is not liable for damages resulting from the advice or assistance, so long as that person did not receive compensation for services in rendering the assistance or advice and did not commit reckless, wanton, or intentional misconduct or gross negligence.18/

5.2.3 Liability of Officers, Directors, and Shareholders

The courts have generally given CERCLA an expansive reading. In particular, courts have held that a "person"19/ who holds an interest in a corporation and has the power to prevent and abate damage from hazardous substances associated with the corporation is liable under CERCLA whether the

---

16/ Iowa Code § 455B.393.

17/ Iowa Code § 455B.393.

18/ Iowa Code § 455B.399.

19/ A "person" is broadly defined to include any individual, partnership, association, public or private corporation or other entity including the United States government, any interstate body, the state and any agency, department or political subdivision of the state.

person is a corporation or individual.[20] Accordingly, a corporate officer, director, or shareholder may be personally liable under CERCLA if the individual had the authority to control practices of the corporation and participated in decisions or had knowledge of activities involving the disposal of hazardous substances.[21]

Notwithstanding the general trend to hold such individuals liable, one federal appellate decision held that shareholders in a corporation may not be liable under CERCLA unless traditional corporate law principles mandate "piercing the veil" of the corporation.[22] The court rejected application of CERCLA liability to a parent corporation. Thus, traditional corporate law principles may provide some protection from Superfund liability.

5.2.4 Parent and Successor Corporate Liability

Like officers, directors, and shareholders, parent and successor corporations have also been held liable under Superfund laws.[23] The general principles

---

20/ Idaho v. Bunker Hill Co., 635 F. Supp. 665, 672 (D. Idaho 1986); see also United States v. Kayser-Roth Corp., 910 F.2d 24, 27 (1st Cir. 1990), cert. denied, 498 U.S. 1084 (1991); NEPACCO, 810 F.2d 726, 743-45 (8th Cir. 1986), cert. denied, 484 U.S. 848 (1987); New York v. Shore Realty Corp., 759 F.2d 1032, 1052 (2d Cir. 1985).

21/ Nurad, Inc. v. William E. Hooper & Sons Co., 966 F.2d 837, 842 (4th Cir. 1992) (declining to impose "operator" liability on sons of corporate president, even though sons were directors and owned corporate stock, because corporate president exercised all operational authority), cert. denied, 113 S. Ct. 377 (1992); Riverside Mkt. Dev. Corp. v. International Bldg. Prods., Inc., 931 F.2d 327, 329-30 (5th Cir. 1991) (affirming summary judgment in favor of 85% shareholder who had not participated in management of company such that he exercised operational control), cert. denied, 112 S. Ct. 636 (1991); Armotek Indus., Inc. v. Freedman, 790 F. Supp. 383, 394-95 (D. Conn. 1992) (granting summary judgment to individual shareholder in "operator" liability case based on lack of personal participation in waste disposal decision); Kelly v. Arco Indus. Corp., 723 F. Supp. 1214, 1217 (W.D. Mich. 1989) (analyzing corporate officer liability).

22/ Joslyn Mfg. Co. v. T. L. James & Co., 893 F.2d 80, 83-85 (5th Cir. 1990), cert. denied, 498 U.S. 1108 (1991); see United States v. TIC Inv. Corp., 1994 Westlaw 608,506 (N.D. Iowa Sept. 19, 1994).

23/ See Smith Land Improvement Corp. v. Celotex Corp., 851 F.2d 86 (3rd Cir. 1988) (successor liability), cert. denied, 488 U.S. 1029 (1989); T & E Indus., Inc. v. Safety Light Corp., 680 F. Supp. 696 (D. N. J. 1988) (successor liability); United States v. McGraw-Edison Co., 718 F. Supp. 154 (W.D.N.Y. 1989) (parent liability); United States v. Nicolet, Inc., 712 F. Supp. 1193 (E.D. Pa. 1989) (parent liability).

governing individual Superfund liability are directly applicable to parent corporation liability due, in large part, to the broad interpretation of the term "person" and the existence of control or knowledge.24/ With regard to successor liability, the acquiring corporation generally does not assume the debts and liabilities of the selling corporation when the corporation purchases the assets only.25/ Where the successor corporation continues the operation of the predecessor, however, courts have tended to impose successor liability as discussed in chapter 16 of this handbook.26/

5.2.5 Lender Liability

As discussed in chapter 16, lenders have also been held liable under Superfund laws when they foreclose on contaminated property or become involved in the management of the property.27/ In 1992, EPA promulgated rules to provide

---

24/ See John S. Boyd Co., Inc. v. Boston Gas Co., 992 F.2d 401, 408 (1st Cir. 1993) (upholding "operator" liability of parent that made essentially all financial decisions, controlled checking account, and handled oil purchases); United States v. Kayser-Roth Corp., 910 F.2d 24, 27 (1st Cir. 1990) (upholding "operator" liability of parent that controlled and participated in managing corporation such that parent was true operator), cert. denied, 498 U.S. 1084 (1991); United States v. TIC Inv. Corp., 1994 Westlaw 608,506 (N.D. Iowa Sept. 19, 1994) (granting summary judgment against parent corporation based on "actual control" over activities of subsidiary); United States v. McGraw-Edison Co., 718 F. Supp. 154, 157 (W.D.N.Y. 1989) (focusing on control over and participation in hazardous waste decisions).

25/ See United States v. Mexico Feed & Seed Co., 980 F.2d 478, 487-90 (8th Cir. 1992); Smith Land & Improvement Corp. v. Celotex Corp., 851 F.2d 86, 91 (3rd Cir. 1988), cert. denied, 488 U.S. 1029 (1989); Anderson v. City of Minnetonka, 1993 Westlaw 95361 (D. Minn. Jan. 27, 1993); Soo Line R. Co. v. B.J. Carney & Co., 797 F. Supp. 1472, 1482 (D. Minn. 1992).

26/ See United States v. Carolina Transformer Co., 978 F.2d 832, 838 (4th Cir. 1992); Anspec v. Johnson Controls, Inc., 922 F.2d 1240, 1246-47 (6th Cir. 1991); Smith Land & Improvement Corp. v. Celotex Corp., 851 F.2d 86, 92 (3rd Cir. 1988), cert denied, 488 U.S. 1029 (1989); United States v. Distler, 741 F. Supp. 637, 642-43 (W.D. Ky. 1990).

27/ See United States v. Fleet Factors Corp., 901 F.2d 1550, 1557-58 (11th Cir. 1990), cert. denied, 498 U.S. 1046 (1991); United States v. Mirabile, 15 Envtl. L. Rep. (Envtl. L. Inst.) 20,994, 20,996 (E.D. Pa. Sept. 4, 1985); United States v. Maryland Bank & Trust Co., 632 F. Supp. 573, 578-80 (D. Md. 1986); Guidice v. BFG Electroplating & Mfg. Co., 732 F. Supp. 556, 561 (W.D. Pa. 1989); see also United States v. McLamb, 5 F.3d 69, 73 (4th Cir. 1993); Ashland Oil, Inc. v. Sonford Prods. Corp., 810 F. Supp. 1057, 1060 (D. Minn. 1993).

guidance in this area, but the D.C. Circuit Court of Appeals struck down these rules.[28/] Congressional amendments to the Superfund law to provide protections to lenders remain under discussion.

While the status of federal law remains in dispute, to address lenders' concerns, the Iowa legislature in 1993 added the following amendment to the Iowa statutory definitions.

> "Person having control over a hazardous substance" does not include a person who holds indicia of ownership in a hazardous condition site, if the person satisfies all of the following:
>
> a. Holds indicia of ownership primarily to protect that person's security interest in the hazardous condition site, where the indicia of ownership was acquired either for the purpose of securing payment of a loan or other indebtedness, or in the course of protecting the security interest. The term "primarily to protect that person's security interest" includes, but is not limited to, ownership interest acquired as a consequence of that person exercising rights as a security interest holder in the hazardous condition site, where the exercise is necessary or appropriate to protect the security interest, to preserve the value of the collateral, or to recover a loan or indebtedness secured by the interest. The person holding indicia of ownership in a hazardous condition site and who requires title or a right to title to the site upon default under the security arrangement, or at, or in lieu of, foreclosure, shall continue to hold the indicia of ownership primarily to protect that person's security interest so long as the subsequent actions of the person with respect to the site are intended to protect the collateral secured by the interest, and demonstrate that the person is seeking to sell or liquidate the secured property rather than holding the property for investment purposes.
>
> b. Does not exhibit managerial control of, or managerial responsibility for, the daily operation of the hazardous condition site through the actual, direct, and continual or recurrent exercise of managerial control over the hazardous condition site in which that person holds a security interest, which managerial divests the borrower, debtor, or obligor of control.

28/ 40 C.F.R. § 300.1100, 57 Fed. Reg. 18,343 (April 29, 1992); Kelley v. EPA, 15 F.3d 1100 (D.C. Cir. 1994).

c. Has taken no subsequent action with respect to the site which causes or exacerbates a release or threatened release of a hazardous substance.29/

5.3 Liability for Response Costs and Natural Resource Damages

CERCLA holds a responsible person liable for federal, state, Indian tribal, local governmental, and private party response costs and damages to natural resources which result from the release or threatened release of hazardous substances.30/ Response costs include all reasonable and necessary costs of investigation, removal, cleanup and remedial actions. Some of these terms, in turn, are defined to include costs to monitor, test, analyze, and confine a release; to clean up, remove, process, and dispose of removed hazardous substances; to relocate residents and businesses; to secure and limit access to a site; and to provide alternative water supplies, among other things.31/ Although the Eighth Circuit had allowed for recovery of attorneys fees under CERCLA, the U.S. Supreme Court in 1994 ruled that CERCLA does not provide for the award of private (non-governmental) litigants' attorneys fees associated with bringing a cost recovery action..32/

Natural resource damages (NRD) generally consist of the residual harm to natural resources after the effects of the environmental cleanup have been taken into account.33/ Federal, state, and tribal trustees may bring claims for natural resource damages under CERCLA, the Clean Water Act, the Oil Pollution Act, and the State of Iowa may assert NRD claims under Iowa statutes.34/ Federal trustees

---

29/ Iowa Code §§ 455B.381(7); 455B.392(7) (providing that under certain circumstances lender may be liable for lesser of cleanup costs or postcleanup fair market value of property); see also First Iowa State Bank v. Iowa DNR, 502 N.W.2d 164 (Iowa 1993).

30/ 42 U.S.C. § 9607(a).

31/ 42 U.S.C. § 9601 (23)-(25).

32/ Key Tronic Corp. v. United States, 114 S. Ct. 1960, 1967-68 (1994); see also United States v. Mexico Feed & Seed Co., 980 F.2d 478, 490 (8th Cir. 1993).

33/ In re Acushnet River & New Bedford Harbor, 712 F. Supp. 1019, 1035 (D. Mass. 1989).

34/ CERCLA, 42 U.S.C. § 9607(a)(4)(C); CWA, 33 U.S.C. § 1321(f)(4); OPA, 33 U.S.C. § 2702(b)(2)(A); Iowa Code § 455B.392.

have adopted damage assessment regulations at 43 C.F.R. Part 11 and 15 C.F.R. Part 990.

Under the Iowa Code § 455B.392, a person having control over a hazardous substance is strictly liable to the state for:

1. Reasonable cleanup costs incurred by the state as a result of the failure of the person to clean up a hazardous substance involved in a hazardous condition caused by that person.

2. Reasonable costs incurred by the state to evacuate people from the area threatened by a hazardous condition caused by the person.

3. Reasonable damages to the state for the injury to, destruction of, or loss of natural resources resulting from a hazardous condition caused by that person including the costs of assessing the injury, destruction, or loss.

4. The excessive and extraordinary cost, excluding salaries, incurred by the DNR in responding at and to the scene of a hazardous condition caused by that person.

5.4 Defenses Under Superfund

The only defenses to liability recognized under CERCLA and the Iowa hazardous conditions statutes are that the release or hazardous condition was caused solely by:

1. An act of God;
2. An act of war; or
3. An act or omission of a third party.[35]

The statutes emphasize that a "third-party" does not include an employee or agent of the defendant or person in the chain of responsibility for the generation, transportation, storage, treatment, or disposal of the hazardous substance.[36] The defense that the release was caused by an act or omission of the third-party is

---

35/ 42 U.S.C. § 9607(b); Iowa Code § 455B.392(3).

36/ Iowa Code § 455B.392(3).

available only if the responsible person establishes the exercise of due care with respect to the hazardous substance concerned. Factors that may be taken into consideration include the characteristics of the hazardous substance in light of all relevant facts and circumstances which defendant knew or should have known and evidence that defendant took precautions against foreseeable acts or omissions and the consequences that foreseeably result from them.[37]

Although federal courts have generally held that only the defenses to liability set forth in the statute are available,[38] some courts have allowed reliance on equitable defenses such as estoppel and government negligence.[39] Under the so-called "useful products" doctrine, courts have declined to impose arranger liability on a manufacturer based solely on the sale of a new product containing hazardous substances as opposed to the sale of hazardous substances to get rid of them.[40] For transporters, a defense to liability under Iowa, but not federal, law exists if the liability arises during the transportation of a hazardous substance, the presence of a hazardous substances was misrepresented to the transporter, and the transporter had no reason to know that the misrepresentation has been made.[41]

37/ Iowa Code § 455B.392(3)(c).

38/ See, e.g., United States v. Stringfellow, 661 F. Supp. 1053, 1062 (C.D. Cal. 1987) and Levin Metals Corp. v. Parr-Richmond Terminal Co., 799 F.2d 1312, 1317 (9th Cir. 1986), cert. denied, 490 U.S. 1098 (1989).

39/ See, e.g., United States v. Mottolo, 695 F. Supp. 615, 626 (D. N.J. 1988); United States v. Hardage, 116 F.R.D. 460, 465 (W.D. Okla. 1987); Violet v. Picillo, 648 F. Supp. 1283, 1294 (D.R.I. 1986).

40/ Florida Power & Light Co. v. Allis Chalmers Corp., 893 F.2d 1313, 1319 (11th Cir. 1990).

41/ Iowa Code § 455B.392(4).

5.5 The National Priorities List and the Superfund Listing Process

CERCLA requires the EPA to establish a National Contigency Plan (NCP) to respond to contaminated sites and list of priorities among sites with releases or threatened releases for the purpose of prioritizing the taking of remedial action.42/ Pursuant to this section and rules adopted by the EPA, the EPA created and updates the National Priorities List (NPL) of Superfund sites on at least an annual basis.

The stated purpose of the NPL is to assist in establishing EPA priorities for taking response actions and for use as a budgetary planning document. Sites are ranked according to the EPA's Hazard Ranking System (HRS) which provides a relative ranking of the actual or potential hazards posed by the site to public health and the environment.43/

Assessment of a site for NPL listing purposes generally consists of four phases:

1. Site discovery;
2. Preliminary assessment;
3. Site inspection; and
4. HRS scoring.

Discovery is a process of identifying potential hazardous waste sites. Sources of information include citizen complaints, hot-line notifications, mandated federal and state site notification and reporting programs, and other government records. Once a potential hazardous waste site is identified, the site is entered into the EPA's Comprehensive Environmental Response, Compensation and Liability Information System (CERCLIS), a listing of all potential hazardous waste sites reported to EPA nationwide.

42/ 42 U.S.C. § 9605; 40 C.F.R. pt. 300.

43/ EPA's HRS evaluates "pathways" such as groundwater, surface water, air, and direct contact by which humans and the environment can be exposed to hazardous substances by assigning numerical values to factors related to each pathway. Priority is given to sites where a public water supply is threatened or contaminated or where drinking water wells have been closed. The HRS process involves both objective and subjective evaluations, and PRPs should involve themselves in it insofar as possible.

The preliminary assessment involves a general review of readily accessible information to characterize the potential hazard and to determine if the site warrants further action. The information gathered during a preliminary assessment (PA) includes a site history, known or alleged hazardous substances present, and potential effects of the contamination on nearby populations and the environment.

A screening site inspection is conducted to define further the extent of the problem and to provide an adequate data base for ranking the site according to its actual or potential hazard. Site-specific data on the hazardous substances present, pollution dispersal pathways, types of receptors, and site management practices are obtained. The site inspection is usually completed within the three to six month period. These rankings are subject to change as more information becomes available although followup evaluations seldom occur without PRP prompting.

If, as a result of the preliminary investigative activities, a site is verified as a hazardous waste site, the site is ranked as to its relative severity against other sites. The HRS scoring model utilizes the information gathered during the preliminary assessment and site inspection. The scores are used to establish priorities among sites and to determine a site's eligibility for federal and/or state Superfund money for response actions.44/ Moneys shall not be used from the Iowa fund for abandoned site cleanup unless the DNR director has made all reasonable efforts to secure voluntary agreement to pay the costs of necessary remedial actions from owners or operators of abandoned or uncontrolled disposal sites or other responsible persons. The DNR is directed to take all reasonable efforts to recover the full amount of moneys expended from the fund through litigation or other means.45/

44/ See Iowa Code § 455B.423.

45/ Iowa Code § 455B.423.

Following the hazard ranking process, a site may be added to EPA's National Priorities List (NPL).46/ The EPA remains the lead agency on all NPL Superfund sites in Iowa.

Iowa has prepared a similar priority list of of 67 contaminated sites in Iowa.47/ The DNR ranks sites based on the following classifications.

Registry Priority Classifications

Class	Action Requirement	Site Condition
"a"	Immediate Action Required	Causing or presenting an imminent danger of irreversible or irreparable damage to the public health or environment.
"b"	Action Required	Significant threat to the environment.
"c"	Action May Be Deferred	Not a significant threat to the environment.
"d"	Requires Continued Management	Site properly closed.
"e"	No Further Action Required	Site properly closed, no evidence of present or potential adverse impact.

The state's registry of abandoned or uncontrolled disposal sites may be the basis to restrict the owner's ability to sell, convey, or transfer title to the site without the DNR's approval.48/ Site owners may contest the listing of a site on the Iowa registry by a contested case proceeding pursuant to the Iowa Administrative Procedure Act.

46/ See 40 C.F.R. pt. 300 App. B.

47/ Iowa Code § 455B.385.

48/ Iowa Code §§ 455B.426, 455B.430.

5.6 Superfund Investigation and Cleanup Process

5.6.1 Remedial Investigation and Feasibility Study (RI/FS)

The Iowa regulations governing the procedures and criteria that the DNR will use to identify PRPs and protect groundwater are codified at Iowa Admin. Code ch. 567-133. Because most Superfund investigations in Iowa take place pursuant to the federal NCP regulations at 40 C.F.R. Part 300, the NCP process is described here.

The purpose of the remedial investigation (RI) is to study groundwater, surface water, soil and/or air which may be affected by the contamination. The RI will provide PRPs and the EPA with information about what contaminants are present at the site and in what quantities; the horizontal and vertical extent of the contamination; how contaminants may have migrated off the site or affected nearby populations or the environment; what risk the contaminants may pose to nearby communities; and how contaminants are affecting the groundwater underneath the site.

The responsible party or parties usually hire an environmental consultant to conduct the RI, with reports submitted to the PRPs and to the EPA for review and approval. At NPL sites where no responsible party has been identified or responsible parties are unwilling or unable to conduct investigations, the EPA hires its own consultant to conduct the RI. Among the activities that typically take place during an RI are sampling of groundwater, surface water, soil or air, or tests to determine whether contaminants are moving from one layer of groundwater to a different groundwater layer.

Often these investigations are performed over time through a phased approach. For example, a limited number of carefully placed groundwater monitoring wells may be installed and sampled. Based on the data gathered, the EPA might require installation of more wells to define more accurately the groundwater movement. Another approach is to break up the site into different operable units -- the groundwater problems may be investigated first, for example, followed by an investigation of the soils or specific areas in which hazardous wastes may have been deposited.

Based on the RI, a study of cleanup alternatives is conducted. Generally, several possible cleanup options are available to address a site or specific portions or units of it, and the feasibility study (FS) is designed to evaluate feasible remedial alternatives and to recommend a remedy which will protect human health and the environment. The FS also estimates the costs and risks of each alternative.

5.6.2 EPA Analysis of Alternatives and Selection of the Remedy

The EPA weighs each option presented in the FS, considering a number of factors, including whether each alternative:

1. Protects public health and the environment;
2. Is in compliance with applicable state and federal laws and regulations (ARARs);49/
3. Would be an effective short-term solution;
4. Would be an effective and permanent long-term solution;
5. Reduces the quantity, toxicity, mobility, or hazardous nature of thorough treatment;
6. Is technically feasible and implementable;
7. Is cost-effective; and
8. Is acceptable to the state and the community.

After screening all the options, the EPA develops a cleanup recommendation, called a Proposed Plan. The agency presents this recommendation at a public

---

49/ ARARs are applicable or relevant and appropriate requirements of federal or state environmental laws. Compliance with ARARs is mandated by the federal National Contingency Plan (NCP). The original plan published by EPA was updated on November 20, 1985 (50 Fed. Reg. 47,912) and March 8, 1990 (55 Fed. Reg. 8,666). Compliance with ARARs frequently determines the cleanup standards applied to a particular site and the objectives the cleanup must achieve since the lead oversight government agency (federal or state) has discretion in deciding what standards a remedy should attain.

meeting and requests comment from the involved community before completing the EPA's final decisionmaking document, called the Record of Decision (ROD).

5.6.3 Remedial Design (RD) and Remedial Action (RA)

After the final decision reflected in the ROD is made, specific engineering plans and specifications are prepared for the cleanup actions. These plans and specifications which provide the construction details are called the Remedial Design (RD), which must be approved before the actual construction and installation of remedial measures to mitigate or immobilize the, or remove contaminated substances (the Remedial Action or RA) can begin.

5.6.4 Long-Term Monitoring and Maintenance

After the cleanup systems are constructed, the site must generally be monitored and the remediation measures operated and maintained (O & M) for the life of the cleanup project. In some cases (a pumpout and treatment of contaminated groundwater, for example), the cleanup itself may take many years. In other cases, removal of the contamination may be addressed quickly, but monitoring of groundwater may be required for several years to corroborate that the removal was complete and effective. Each site is different, but most have monitoring and maintenance requirements to assure that the cleanup is effective over the long term. Unfortunately, regulatory agencies are most reluctant to acknowledge that remediation has achieved its anticipated objectives, that O & M may cease, and that the site may be removed from the Superfund list.

5.7 Contribution and Apportionment of Liability

CERCLA case law provides that any person held jointly or severally liable for response costs and natural resource damages may be entitled to have the trier of fact apportion liability among the parties. The burden is on each defendant to show

how that person's liability should be apportioned, and the court may reduce the amount of damages in proportion to the amount of liability so apportioned.50/ Although not codified in CERCLA, the following factors proposed by then-Senator Gore (the "Gore factors") frequently are considered in apportioning liability:

1. The extent to which that party's contribution to the release of a hazardous substance can be distinguished;

2. The amount of hazardous substance involved;

3. The degree of toxicity of the hazardous substance involved;

4. The degree of involvement of and care exercised by the party in manufacturing, treating, transporting, and disposing of the hazardous substance;

5. The degree of cooperation by the party with federal, state, or local officials to prevent any harm to the public health or the environment; and

6. Knowledge by the party of the hazardous nature of the substance.

Indemnification agreements, hold harmless agreements, or conveyances of property interest do not avoid a person's liability under CERCLA or the Iowa hazardous conditions statutes. However, such documents may be useful in resolving anticipated future allocation problems. Moreover, they are legally enforceable as between responsible parties, and thus may eliminate future disputes between them.51/

50/ See Gopher Oil Co. v. Union Oil Co., 955 F.2d 519, 526 (8th Cir. 1992) ("Allocation of liability under CERCLA . . . is an equitable determination . . ..").

51/ See chapter 16 of this handbook, which deals with environmental considerations in real estate and business transactions.

5.8 Government Enforcement Actions

While a site is being scored and evaluated, the EPA or the DNR generally engages in a responsible party search, often by use of a CERCLA § 104 request for information. These information requests seek written responses to questions relating to possible association with and knowledge of a site. The EPA or the DNR then may seek to commence negotiations with the PRPs in an effort to reach a consent order to undertake the RI/FS work and the RD/RA work. If a consent order cannot be reached, the EPA may issue a unilateral administrative order pursuant to CERCLA § 106 directing the PRP or PRPs to undertake the work.52/ As an alternative, the EPA may undertake the work itself and seek reimbursement of its response costs from the PRP or PRPs in litigation.

A person shall not refuse entry or access to, or harass or obstruct an authorized representative of the DNR who seeks entry or access for the purpose of investigating or responding to a hazardous condition.53/ Upon a showing of probable cause, the DNR may obtain a search warrant to investigate or respond to a hazardous condition.

As discussed above, the Iowa hazardous conditions statutes hold a responsible person liable for all reasonable response costs incurred by the state. Liability to the state for reasonable cleanup expenses incurred by the state is a lien upon the real estate that attaches at the time the state incurs the expenses.54/ When any hazardous condition exists, the DNR director may remove or provide for the removal and disposal of the hazardous substance at any time, unless the director determines such removal will be properly and promptly accomplished by the owner or operator of the vessel, vehicle, container, pipeline, or other facility.55/

52/ See Dico, Inc. v. Diamond, 821 F. Supp. 562 (S.D. Iowa 1993).

53/ Iowa Code § 455B.394.

54/ Iowa Code §§ 455B.392(7)(d), 455B.396.

55/ Iowa Code § 455B.387.

If the DNR director determines that an emergency exists respecting any matter affecting or likely to affect the public health, the director may issue an order necessary to terminate the emergency without notice and without hearing.[56] The DNR director also may request that the attorney general institute legal proceedings to enforce the Iowa hazardous conditions statutes, and upon such a request, the attorney general shall institute any legal proceedings necessary to obtain compliance with these statutes.[57]

56/ Iowa Code § 455B.388; see also Cota v. Iowa Envtl. Protection Comm'n, 490 N.W.2d 549 (Iowa 1992); Pruess Elevator, Inc. v. Iowa DNR, 477 N.W.2d 675 (Iowa 1991).

57/ Iowa Code §§ 455B.388, 455B.391.

CHAPTER SIX

HAZARDOUS WASTE MANAGEMENT

6.0 Relationship of Hazardous Waste Management in Iowa to the Federal Resource Conservation and Recovery Act (RCRA)

In response to public concern over improper handling and disposal of chemical wastes, the United States Congress enacted the Resource Conservation and Recovery Act (RCRA) in 1976.[1/] RCRA authorizes the United States Environmental Protection Agency (EPA) to develop and oversee a comprehensive nationwide program for the management of hazardous waste from its generation to ultimate disposal -- so-called "cradle to grave" regulation.[2/] RCRA provides that states may develop their own statutory framework within which to implement the requirements of RCRA.[3/] Congress subsequently added to RCRA the Hazardous and Solid Waste Amendments of 1984[4/] which were primarily concerned with placing stringent limitations on land disposal of hazardous wastes and regulating underground storage tanks.

Under RCRA, EPA promulgated regulations which became effective in November 1980. These regulations provide:

1. A definition of hazardous waste, including lists of specific types of hazardous waste;[5/]

2. Handling requirements for generators and transporters of hazardous waste;[6/]

1/ Pub. L. No. 94-580, 90 Stat. 2796 (1976). Under 42 U.S.C. ch. 82, RCRA regulates both hazardous waste under Subchapter III (Subtitle C) of the Act and solid or non-hazardous waste under Subchapter IV (Subtitle D). RCRA also regulates underground storage tanks, 42 U.S.C. § 6991, and promotes resource recovery and reuse, 42 U.S.C. § 6962.

2/ See 42 U.S.C. §§ 6901-6991; see also City of Chicago v. Environmental Defense Fund, 114 S. Ct. 1588, 1590 (1994).

3/ 42 U.S.C. § 6926.

4/ Pub. L. No. 98-616, 98 Stat. 2224 (1984).

5/ 40 C.F.R. pt. 261.

6/ 40 C.F.R. pts. 262, 263, and 266.

3. A "cradle-to-grave" manifest system to track hazardous waste from the site of its generation to its final disposal;7/

4. Permit requirements for facilities that treat, store, or dispose of hazardous waste;8/ and

5. Requirements for state hazardous waste programs.9/

RCRA provided that a state may apply to the EPA for authorization to operate its own program in lieu of the RCRA program. Iowa initially adopted a hazardous waste management program to operate in lieu of the federal program, but returned this authority to EPA in 1985. While Iowa has adopted certain hazardous waste laws that remain on the books, EPA administers most hazardous waste regulations in Iowa.

6.1 Waste Evaluation: Is the Waste Regulated?

6.1.1 Overview

Hazardous waste is a subset of "solid waste," which is defined in 40 C.F.R. Part 261.2. A person who generates solid waste must determine if that waste is a "hazardous waste."10/

The statutory definition of a "hazardous waste" is as follows:11/

> [A] solid waste, or combination of solid wastes, which because of its quantity, concentration, or physical, chemical, or infectious characteristics may -- (a) cause or significantly contribute to an increase in mortality or an increase in

---

7/ 40 C.F.R. § 262.20.

8/ 40 C.F.R. pts. 264-266, 270.

9/ 40 C.F. R. pt. 271. Jurisdiction over hazardous waste on Native American lands has been more controversial. See Note, Regulatory Jurisdiction over Non-Indian Hazardous Waste in Indian Country, 72 Iowa L. Rev. 1091 (1987).

10/ 40 C.F.R. § 262.11. The leading treatise on hazardous waste law discusses the definition of hazardous waste in detail and was written by University of Iowa Professor John Mark Stensvaag, Hazardous Waste Law and Practice (1994).

11/ 42 U.S.C. § 6903(5); see also 40 C.F.R. pt. 261.

serious irreversible, or incapacitating reversible, illness; or (b) pose a substantial present or potential hazard to human health or the environment when improperly treated, stored, transported, or disposed of, or otherwise managed.

As a practical matter, the first step in evaluating waste is to identify or inventory a facility's waste. Sources of information include purchasing departments, production maintenance personnel, inspections of loading docks and storage areas, material safety data sheets (MSDSs), and shipping documents. Depending upon the amount of information available, it may or may not be necessary to have specific wastes analyzed by a laboratory.

6.1.2 Exempt Wastes and Containers

Certain solid wastes are exempt from hazardous waste regulation:

1. Normal refuse from households, motels, and hotels including garbage, trash, and sanitary wastes in septic tanks;

2. Solid wastes generated by agricultural crops or livestock which are returned to the soil as fertilizers;

3. Mining overburden returned to the mine site;

4. Certain ash waste from the combustion of coal or other fossil fuels;[12]

5. Drilling fluids, produced waters, and other wastes associated with oil and gas production;

6. Certain waste containing trivalent (not hexavalent) chromium;

7. Certain waste from the extraction, beneficiation, and processing of ores and minerals;

8. Cement kiln dust waste, except in some cases where hazardous waste is processed in the kiln; and

---

12/ 40 C.F.R. § 261.4; see also City of Chicago v. Environmental Defense Fund, 114 S. Ct. 1588 (1994) (incinerator ash is not excluded from regulation as "hazardous waste").

9. Certain petroleum-contaminated media that are subject to corrective action requirements under underground storage tank regulations;

10. Waste water, soil, or air samples collected for the purpose of determining characteristics or composition.

This is not a complete listing of exempt wastes; consult the rules for other categories.

Empty containers and liners that have held a hazardous waste, except containers that have held a compressed gas or an acute hazardous waste,13/ are exempt if all the removable wastes have been removed from the container, using common practices, and one of the following conditions is met:

1. No more than one inch of residue remains on the bottom;

2. No more than 3.0 percent (by weight) waste remains inside the container having a total capacity of 110 gallons or less; or

3. No more than 0.3 percent (by weight) waste remains inside the container having a total capacity of more than 110 gallons.14/

A container that has held an acute hazardous waste is deemed empty if:

1. The container has been triple rinsed;

2. It has been cleaned by a method demonstrated or recognized to obtain equivalent removal; or

3. In the case of a container, the inner liner has prevented contact and has been removed.15/

A container that has held a compressed gas that is a hazardous waste is deemed empty when the pressure in the container approaches atmospheric pressure.16/ A farmer disposing of waste pesticides from his or her own use is

---

13/ 40 C.F.R. § 261.7.

14/ 40 C.F.R. § 261.7(b)(1).

15/ 40 C.F.R. § 261.7(b)(3).

16/ 40 C.F.R. § 261.7(b)(2).

exempt provided that emptied pesticide containers are triple-rinsed and pesticide residues are disposed of on the farmer's own farm consistent with disposal instructions on the pesticide label.17/

6.1.3 Listed Hazardous Wastes

The primary methods of determining whether a waste is hazardous are: (1) whether it is listed in the rules as a hazardous waste; and (2) whether it exhibits hazardous characteristics such as ignitability, corrosivity, reactivity, or toxicity.

The lists of hazardous wastes cover dozens of pages.18/ Each listed waste is assigned a hazardous waste number and, in many cases, a hazard code which must be used in complying with disclosure, recordkeeping, and reporting requirements. Hazardous wastes are listed as those from (1) "non-specific sources;"19/ (2) "specific sources;"20/ and (3) discarded commercial chemical products by chemical name listed alphabetically.21/ The waste from nonspecific sources are associated with particular processes such as electroplating, pesticide manufacturing, wood preservative manufacturing, and petroleum refining. All lists must be consulted to evaluate specific wastes and waste streams.

6.1.4 Waste Exhibiting Hazardous Characteristics

A waste which is not classified as a listed hazardous waste is nonetheless a hazardous waste if it exhibits the characteristics of ignitability, corrosivity, reactivity, or toxicity.22/

17/ 40 C.F.R. § 262.70.

18/ 40 C.F.R. §§ 261.30-261.35.

19/ 40 C.F.R. § 261.31.

20/ 40 C.F.R. § 261.32.

21/ 40 C.F.R. § 261.33.

22/ 40 C.F.R. §§ 261.20-261.24.

• **Ignitability**. A liquid is ignitable if it has a flash point below 140 degrees fahrenheit. This can be determined from the MSDS or laboratory analysis report. Many solvents are ignitable. A solid waste is also ignitable if it can spontaneously catch fire and burn so vigorously and persistently that it represents a hazard. Ignitable wastes carry the EPA Hazardous Waste Number of D001.[23]

• **Corrosivity**. This category of hazardous wastes generally consist of strong acids or bases. Any water-based waste having a pH of less than 2.0 or equal to or greater than 12.5 is corrosive. A liquid able to corrode steel at a rate of 0.25 inches per year is also corrosive. Corrosive wastes carry the EPA Hazardous Waste Number of D002.[24]

• **Reactivity**. Unstable or explosive wastes, wastes which react violently when brought in contact with water, and wastes which can release toxic vapors such as hydrogen cyanide or hydrogen sulfide are considered reactive. Reactive wastes carry the EPA Hazardous Waste Number of D003.[25]

• **Toxicity (by TCLP test)**. A waste that releases certain amounts of specified organic and inorganic substances in a laboratory extraction procedure is called "toxic" and is hazardous. EPA uses the Toxicity Characteristic Leaching Procedure (TCLP) testing method for toxicity. The rule lists the maximum concentrations allowed for the relevant substances.[26] Wastes commonly requiring TCLP toxicity testing include paints, inks, sludges, and photodeveloping solutions. These wastes carry the EPA Hazardous Waste number corresponding to the chemical causing it to be hazardous.

23/ 40 C.F.R. § 261.21.

24/ 40 C.F.R. § 261.22.

25/ 40 C.F.R. § 261.23.

26/ 40 C.F.R. § 261.24.

6.1.5 Waste Mixtures and Wastes Derived From Hazardous Wastes

EPA promulgated regulations in 1980 which reflected presumptions that: (1) a mixture of non-hazardous and listed hazardous wastes is generally deemed hazardous (the "mixture rule"); and solid waste derived from the treatment or disposal of a listed hazardous waste is itself hazardous waste (the "derived from rule").27/ Eleven years later, the regulations were declared invalid due to procedural irregularities in their promulgation.28/ After that decision, EPA rescinded the rules and replaced them with virtually identical temporary rules.29/ The agency then announced proposed new rules which were themselves withdrawn in October 1992.30/ The interim rules (in essence, the 1980 rules) continue in effect while EPA contemplates new rules.

6.2 Generator Requirements

6.2.1 Generator Identification Number

Any generator who treats, stores, disposes, or arranges for the transportation of any hazardous waste must obtain a generator identification number from the EPA.31/ This is accomplished by completion and submission to EPA of the EPA Notification of Hazardous Waste Activity Form. If a generator's operations are on contiguous property, it may use a single EPA generator identification number for all operations even if they are in different buildings. However, if a facility operates with buildings which do not share a common site, even separated by a public

---

27/ 40 C.F.R. § 261.3. Several narrow exceptions to these presumptions are described in the regulation.

28/ Shell Oil Co. v. EPA, 950 F.2d 741 (D.C. Cir. 1991). A subsequent decision held the "mixture and derived from" rules were invalid from the date of their promulgation. United States v. Goodner Bros. Aircraft, 966 F.2d 380 (8th Cir. 1992), cert. denied, 122 L. Ed.2d 123 (1993).

29/ 57 Fed. Reg. 7,628 (1992); 57 Fed. Reg. 7,636 (Mar. 3, 1992).

30/ 57 Fed. Reg. 21,450 (proposed May 20, 1992); 57 Fed. Reg. 49,280 (withdrawn Oct. 30, 1992); see also Mobil Oil Corp. v. EPA, 35 F.3d 579 (D.C. Cir. 1994).

31/ 40 C.F.R. § 262.12. The only exception is for conditionally exempt small quantity generators.

roadway, separate identification numbers must be obtained for each site. The EPA identification number is site-specific and thus will not change unless the facility moves to another location. The EPA must be notified if the name of the facility or the owner changes although the EPA identification number will not.

6.2.2 Generator Size and Categories

The level of regulation which pertains to generators of hazardous waste depends on the amount of hazardous waste generated or accumulated. If hazardous waste generation is seasonal or irregular, the EPA deems one monthly exceedance of the threshold level to raise the generator into the higher category.32/

A large quantity generator (LQG) generates more than 1,000 kilograms (2,200 pounds) of hazardous waste per month. This is approximately five 55 gallon drums. A small quantity (SQG) generator generates 100 kilograms (220 pounds) to 1,000 kilograms (2,200 pounds) of hazardous waste per month. This amounts to approximately one-half to five drums per month. A conditionally exempt small quantity generator (CESQG) generates less than 100 kilograms or 220 pounds per month (approximately one-half drum per month).33/

6.2.3 Reporting Requirements

Large quantity generators must submit to EPA a biennial report that includes the following information:

1. EPA ID numbers, name and address of generator, transporter, and off-site disposal facility, calendar year covered by report;

2. A list of hazardous wastes generated, the hazardous waste numbers if listed wastes, the physical state, and the source or process from which each waste was generated;

32/ See 40 C.F.R. §§ 261.5, 262.34(d), 262.44.

33/ 40 C.F.R. § 261.5.

3. A description of efforts undertaken to reduce the volume and toxicity of waste generated;

4. A description of the changes in volume and toxicity of waste actually achieved in comparison to prior years; and

5. A signature by the generator or authorized representative.[34]

Occasionally, a person who is not regularly and currently generating hazardous waste seeks a one-time disposal of a hazardous waste. One-time disposers who generate hazardous waste must obtain an EPA identification number.

6.2.4 Accumulation and Storage of Hazardous Wastes

A generator may accumulate hazardous waste on-site without a permit or without having interim status if all accumulated hazardous waste is, within ninety days of the "accumulation start date," shipped off-site to a designated facility.[35] The "accumulation start date" begins when the generator initiates accumulation in a container or tank.[36]

A generator who accumulates hazardous waste for more than ninety days is an operator of a storage facility and is subject to the requirements governing storage facilities, unless an extension to the ninety day period has been granted.[37] An extension of up to thirty days may be granted at the discretion of the EPA on a case-by-case basis for unforeseen, temporary, and uncontrollable circumstances.[38]

A small quantity generator who does not have a permit or interim status may accumulate hazardous waste on-site for up to 180 days if the quantity of waste accumulated never exceeds 6,000 kilograms and if the generator meets specified

---

34/ 40 C.F.R. § 262.41.

35/ 40 C.F.R. § 262.34.

36/ 40 C.F.R. § 262.34(a)(2).

37/ 40 C.F.R. § 262.34(b).

38/ 40 C.F.R. § 262.34(b).

management requirements.[39] If the facility receiving the hazardous waste is more than 200 miles from the generation site, a small quantity generator may store waste for up to 270 days.[40]

A conditionally exempt small quantity generator may store up to 1,000 kilograms (2,200 pounds) of hazardous waste indefinitely. Once 1,000 kilograms have accumulated, the waste must be shipped off-site within 180 days.[41] If the receiving facility is more than 200 miles from the generation site, the conditionally exempt small quantity generator may store waste up to 270 days.

6.2.5 Hazardous Waste Storage Requirements

Hazardous waste must be stored safely in an appropriate container or tank.[42] The type of container depends primarily on the type of waste generated. Containers must be made of sturdy leakproof materials and must meet the Department of Transportation specifications for materials and construction. Containers holding hazardous waste must be closed during storage, except to add or remove waste. Containers must be marked with the words "Hazardous Waste."[43] The container must be labeled in a manner that clearly identifies or describes its contents to employees and emergency personnel. Each container must also have the "accumulation start date" marked on it.[44] Containers must be inspected at least weekly for leaks and for deterioration caused by corrosion or other factors. Generators also may use tanks to accumulate hazardous waste, but containers are

---

39/ 40 C.F.R. § 262.34(d).

40/ 40 C.F.R. § 262.34(e).

41/ 40 C.F.R. §§ 261.5(g)(2), 262.34(d).

42/ 40 C.F.R. §§ 262.34(a)(1)(i), 265.170-265.177.

43/ 40 C.F.R. §§ 262.32, 262.34(a)(3).

44/ 40 C.F.R. § 262.34(a)(2).

more common because the standards applicable to tanks are more complicated and stringent.[45]

Many generators produce one or more hazardous waste streams which accumulate very slowly. In such cases, the storage time limits of 90 or 180 days would be difficult to comply with since the container may only be partially full when the limit is up. The regulations allow for longer accumulation times under what is referred to "satellite accumulation."

6.2.6 Satellite Accumulation of Hazardous Waste

Under certain conditions, the satellite accumulation provisions allow a generator to completely fill a container with waste before the storage time clock begins running. This means that a small quantity generator would have 180 days from the date the container becomes full to ship the waste off-site.

Containers (tanks are ineligible) used to accumulate hazardous waste must meet the following requirements in order to qualify as "satellite accumulation:"

1. Containers stored at or near the place where the waste is generated under control of the process supervisor;
2. No more than 55 gallons of non-acutely hazardous waste is stored at the site;
3. The container is labeled "Hazardous Waste" and complies with requirements relating to proper use of containers;
4. The date that the container becomes full is marked on it; and
5. The container is moved to a central storage location within three days of becoming full.[46]

A generator may have more than one satellite accumulation area if hazardous waste is generated in more than one place in the facility.

---

[45] 40 C.F.R. §§ 262.34(a)(1)(ii); see also 40 C.F.R. pt. 265, subpt. J.

[46] 40 C.F.R. § 262.34(c).

6.2.7 Arranging for Transportation and Disposal of Hazardous Wastes

The EPA cautions each generator that it is responsible for ensuring that the hazardous waste transporter it uses is complying with the requirements of state and federal law.[47] Generators are urged to research qualifications and experience of transporters, review their registrations to deliver hazardous waste in the destination state and possession of a U.S. Department of Transportation (DOT) hazardous waste hauler's license, and ensure that the transportation company's drivers are qualified under the Motor Carrier Safety Act.[48] A generator should request that the transporter demonstrate possession of adequate liability insurance.

Similarly, because the generator is ultimately responsible for its waste, it is important to choose a recycling, treatment, or disposal facility that manages hazardous waste in accordance with state and federal regulations. Proper management, especially in final disposal, will go a long way toward protecting generators from liability under state and federal Superfund acts for remediation costs and environmental damages resulting from mismanagement. The best way to assess how waste will be managed is to have a representative visit the intended disposal facility. If this is prohibitive, contact with the regulatory agency in the state where the facility is located will often provide helpful information. In addition, many consultants and environmental lawyers can obtain information.

Before arranging hazardous waste shipments, a generator should obtain a copy of the treatment or disposal facility's certificate of insurance and request that the generator be named on a separate certificate covering its liability in the event that the disposal facility goes bankrupt. In addition, the generator should obtain a sample contract used by the facility that governs its services and liabilities. A generator must know where its waste is going, how it is being managed, and the

---

47/ See also 40 C.F.R. §§ 262.30-262.33 (pre-transport requirements for generators).

48/ 49 U.S.C. § 5109 (shipper may offer hazardous material for transport only if carrier has DOT safety permit); 40 C.F.R. § 262.12(c) (carrier must have EPA identification number).

disposition of any residues, ash, and empty drums. Finally, generators should obtain a certificate of treatment or disposal from the destination company.

6.2.8 Personnel Training

There are no express EPA requirements for CESQGs to train employees in proper hazardous waste management, although employees should be familiar with safe waste handling procedures and proper emergency response procedures. Also, certain counties may have adopted additional requirements. SQGs must ensure that all employees are thoroughly familiar with proper waste handling and emergency procedures, relevant to their responsibilities during normal facility operations and emergencies.[49] LQGs must comply with the personnel training requirements applicable to TSD facilities.[50]

Training should be documented, which means that records must be maintained including dates and content of training for each employee. Training should cover information on the proper storage of containers, labeling, container handling, hazardous waste accumulation rules (storage volumes, time limits, aisle space, and the requirement that containers be closed), preparing containers for shipment, and manifesting. In addition, employees should be trained in proper response procedures in the event of a fire, spill, or other emergency.

The requirements for LQGs include classroom instruction or on-the-job training that teaches employees to perform their duties in a way that ensures the facility's compliance with the regulations. The program must be directed by a person trained in hazardous waste management procedures. Minimum program requirements are outlined in the regulations.[51] Records must be maintained for each employee and for current personnel until closure of the facility. Records of

---

49/ 40 C.F.R. § 262.34(d)(5)(iii).

50/ 40 C.F.R. §§ 262.34(a)(4), 265.16.

51/ 40 C.F.R. § 265.16.

former employees must be kept at least three years from the date that the employee last worked at the facility.

6.2.9 Emergency Planning and Response Requirements

All generators of hazardous waste are required to manage their waste to minimize the chance of accident and to complete emergency planning requirements that describe personnel responses to emergencies. Emergency planning requirements are based on generator category.

SQGs are required to identify an emergency coordinator who must either be on the generator's premises or must be available to respond to an emergency by reaching the premises within a short period of time.52/ SQGs must post emergency notification information, location of fire extinguishers and spill control equipment, fire alarms, and similar information.53/ They must ensure and document that employees are thoroughly familiar with the proper waste handling and emergency procedures relevant to their responsibilities and must set up procedures which would minimize the possibility of an accident.

If a hazardous waste spill occurs which may pollute the air, soil or water, the emergency coordinator or person in control must evaluate the situation, contain and clean up the hazardous waste and contaminated material, and notify the National Response Center.54/ Emergency response requirements for LQGs are more comprehensive and require particular equipment to be on site,55/ preparation of a contingency plan,56/ and the appointment of an emergency coordinator.57/

52/ 40 C.F.R. §§ 262.34(d)(5).

53/ 40 C.F.R. § 262.34(d)(5).

54/ 40 C.F.R. §§ 262.34(a)(4), 262.34(d)(5)(iv) (for SQGs), 265.56 (for LQGs). See chapter 8 of this handbook for further discussion of release reporting requirements.

55/ 40 C.F.R. §§ 262.34(a)(4), 265.32.

56/ 40 C.F.R. §§ 262.34(a)(4), 265.51.

57/ 40 C.F.R. §§ 262.34(a)(4), 265.55.

Emergency procedures and post-emergency requirements for large quantity generators are prescribed by the rules.[58]

6.2.10 Generator Fees

The Iowa DNR oversees a system of fees paid by hazardous waste generators pursuant to Iowa Code § 455B.424 and Iowa Administrative Code ch. 567-149. These fees apply to generators and hazardous waste disposal facilities.

6.3 Transporter and Manifesting Requirements

The U.S. Department of Transportation has adopted regulations governing the transportation of hazardous materials at 49 C.F.R. Parts 171-179. EPA has adopted these regulations by reference with regard to hazardous waste transportation.[59] In addition, all hazardous waste transporters must have an EPA identification number.[60] Iowa's motor carrier safety rules and hazardous material transportation rules are codified at Iowa Admin. Code ch. 761-520.

A transporter may not accept hazardous waste from a generator unless it is accompanied by a manifest signed by the generator,[61] and the transporter must sign and date the manifest acknowledging acceptance of hazardous waste from the generator before transporting it. The transporter must return a signed copy to the generator before leaving the generator's property. The manifest must accompany the shipment in an accessible location. A transporter who delivers a hazardous waste to another transporter designated on the manifest or to the designated facility must obtain the signature of the accepting transporter or facility and the date of

---

58/ 40 C.F.R. §§ 262.34(a)(4), 265.56.

59/ 40 C.F.R. § 263.10.

60/ 40 C.F.R. § 263.11; see also 49 U.S.C. § 5109 (motor carrier must obtain U.S. DOT safety permit to transport hazardous material).

61/ 40 C.F.R. § 263.20; see also 49 U.S.C. § 5110; State v. Presto-X Co., 417 N.W.2d 199 (Iowa 1987).

delivery.62/ If the transporter cannot so deliver the waste, it must contact the generator.63/ The transporter must keep a copy of the manifest signed by the generator, by the transporter, and by the owner or operator of the designated facility for three years from the date that the waste was accepted by the initial transporter.64/ If in the course of transportation a container is discovered to be broken or leaking, the transporter must take appropriate immediate action, notify state and federal agencies, and recover the hazardous waste.65/

6.4 Treatment, Storage, and Disposal (TSD) Facility Requirements

6.4.1 General Requirements; TSD Permitting

Owners and operators of facilities that treat, store, or dispose of hazardous wastes (TSD facilities) must apply for an identification number from the EPA.66/ A TSD facility must notify the EPA if arrangements have been made to receive hazardous wastes from a foreign source.67/ Moreover, an owner or operator of a TSD facility may not accept a shipment of wastes which it is not allowed to manage under its permit. When an owner or operator receives hazardous wastes from an off-site source, it is required to inform the generator in writing that it has the appropriate permits for the waste.68/ An owner or operator that transfers ownership of a TSD facility during its operating life or post-closure care period is required to notify the new owner of the regulations governing management of the facility and all permit requirements. Finally, the owner or operator must meet stringent financial responsibility requirements to cover sudden and non-sudden occurrences

---

62/ 40 C.F.R. § 263.20(d).

63/ 40 C.F.R. § 263.21(b).

64/ 40 C.F.R. § 263.22.

65/ 40 C.F.R. §§ 263.30-263.31.

66/ 40 C.F.R. § 264.11.

67/ 40 C.F.R. § 264.12.

68/ 40 C.F.R. § 264.12(b).

arising from operation of the facility, which may cause personal injury or property damage to third parties.69/

A hazardous waste facility permit issued by the EPA is required for any facility that:

1. Treats, stores, or disposes of hazardous waste;

2. Establishes, constructs, operates, or closes a hazardous waste facility;

3. Expands, increases, or modifies a process that results in new or increased capabilities of a permitted hazardous waste facility; or

4. Operates a permitted hazardous waste facility or part of a facility that has been changed, added to, or extended, or that has new or increased capabilities.70/

Excluded from the permitting requirement are generators who store for less than applicable time limits, farmers who use pesticides, totally enclosed treatment facilities, elementary neutralization units or wastewater treatment units, storage by transporters for ten days or less, persons adding absorbent material to waste in a container, and spill cleanups.71/

Certain types of qualifying facilities are deemed to have obtained a hazardous waste facility permit without making application for it, which is sometimes referred to as "permitting by rule."72/ Qualifying facilities include ocean disposal vessels, publicly owned treatment works if the operator complies with its National Pollutant Discharge Elimination System (NPDES) permit, and certain injection wells.

The application for a hazardous waste facility permit consists of Part A and Part B, which are described in detail in the regulations.73/ Owners or operators of

---

69/ 40 C.F.R. §§ 264.140-264.151.

70/ 40 C.F.R. pt. 270.

71/ 40 C.F.R. § 270.1(c).

72/ 40 C.F.R. § 270.60.

73/ 40 C.F.R. §§ 270.10-270.26.

existing covered facilities already should have submitted Part A of the application to obtain interim status,[74] and, if requested by the EPA, Part B. A person who proposes to construct a new TSD facility must submit Part A and Part B at least 180 days before the planned commencement of construction.[75] Applications must contain certification by the applicant and engineer.

6.4.2 TSD Safety and Security

The rules governing TSD facilities set forth numerous safety and security requirements. Owners and operators are required to prevent unauthorized entry into a hazardous waste facility by use of a 24-hour surveillance system or a barrier that completely surrounds the active portion of the facility with a means to control entry at all times.[76] There are also sign-posting requirements for TSD facilities. Although the rules provide for exemptions from security requirements, the EPA has indicated that it does not expect to grant many exemptions. Adequate aisle space must be maintained.[77]

6.4.3 TSD Inspection Requirements

Owners and operators of TSD facilities must inspect the facility and all operating, monitoring, and safety equipment for signs of deterioration and malfunctions.[78] They must develop and follow a written schedule for these inspections and identify types of potential problems to look for. Written inspection logs must be maintained for at least three years which include the date and time of the inspection, name of the inspector, observations, and the nature and date of repairs or remedial action.

74/ 40 C.F.R. § 270.10.

75/ 40 C.F.R. § 270.10(f).

76/ 40 C.F.R. § 264.14.

77/ 40 C.F.R. § 264.35.

78/ 40 C.F.R. § 264.15.

6.4.4 TSD Personnel Training

Owners and operators of TSD facilities must provide for classroom instruction or on-the-job training for facility personnel. Facility personnel are required to complete a prescribed education program within six months after their assignment or reassignment to a new position within the facility.79/ They also must participate in an annual review of this initial training. Personnel records documenting the training provided to each employee must be maintained by the facility for at least three years from the date the employee last worked at the facility.

6.4.5 TSD Emergency Procedures

The owner or operator of a TSD facility must maintain a contingency plan that contains prescribed information designed to minimize hazards from fires, explosions, or any releases of hazardous waste or hazardous waste constituents to land, air, or water.80/ The TSD facility is also responsible for making arrangements with local authorities in the case of an emergency. At least one employee on the facility premises or on call must be responsible for coordinating emergency response measures in the event of a release at the facility. The emergency coordinator is required to activate alarms and to notify facility personnel and federal, state, and local agencies with designated response roles. The coordinator must review materials and assess possible hazards. The emergency coordinator is also responsible for containment measures, facility monitoring, and cleanup.

79/ 40 C.F.R. § 264.16; see also 29 C.F.R. § 1910.120(p) (OSHA HAZWOPER training); 49 C.F.R. § 172.101 (DOT training for personnel involved in shipping regulated hazardous materials). In September 1994, the Iowa Labor Services Division proposed additional HAZWOPER rules to be codified at Iowa Admin. Code ch. 347-10.

80/ 40 C.F.R. §§ 264.50-264.56.

6.4.6 TSD Waste Analysis

Before treating, storing, or disposing of any hazardous waste, the owner or operator of a TSD facility is required to obtain a detailed chemical and physical analysis of representative samples of incoming wastes and must develop and follow a written waste analysis plan.[81] The rules also provide general requirements for treating, storing and disposing of ignitable, reactive, or incompatible wastes.[82]

6.4.7 TSD Recordkeeping and Reporting

As is the case with generators, the owner or operator of a TSD facility is subject to manifesting[83] and operating recordkeeping requirements.[84] Owners and operators must prepare and submit a biennial report covering, among other things: (a) the identification of all generators from whom waste was received; (b) a description and volume of each waste handled by the facility; (c) the method of treatment, storage, or disposal; and (d) an updated estimate of closure costs.[85] In addition, an incident report of unmanifested wastes received, fires, releases or explosions, and closures is required.

6.4.8 Groundwater Protection at Hazardous Waste TSDs

A groundwater monitoring and detection program is required for all facilities that treat, store, or dispose of hazardous waste in surface impoundments, waste piles, land treatment units, or landfills.[86] The groundwater protection requirements establish a three-stage program to detect, evaluate and, if necessary, correct groundwater contamination during the active life of the facility. This

---

81/ 40 C.F.R. § 264.13.

82/ 40 C.F.R. § 264.17.

83/ 40 C.F.R. §§ 264.70-264.72.

84/ 40 C.F.R. §§ 264.73-264.77.

85/ 40 C.F.R. § 264.75.

86/ 40 C.F.R. §§ 264.90-264.94.

includes the closure period plus any compliance period established in the permit. Groundwater assessment monitoring information must be submitted to EPA.

6.4.9 TSD Design and Operating Requirements

The TSD facility standards include specific design and operating requirements governing the storage or treatment of hazardous wastes in containers, tanks, surface impoundments, waste piles, land treatment and landfills, and incinerators.87/ For example, surface impoundments and landfills are required to have a double liner system.

6.4.10 TSD Closure Requirements; Post-Closure Care; and Financial Assurance Requirements

Specific procedures and standards are prescribed for the closure of a hazardous waste TSD facility and for post-closure care of the property.88/ Specific financial responsibility requirements set forth methods that may be used by an owner or operator to ensure that funds will be available for closure and post-closure care.89/

Within 60 days after closure, the owner or operation of a TSD facility must: (1) record a notation on the deed of the facility property that it was used to manage hazardous waste and that future land use is restricted; and (2) inform the local zoning authority of the type, location, and quantity of hazardous waste disposal in each cell of the facility.90/

87/ 40 C.F.R. §§ 264.170-264.351.

88/ 40 C.F.R. §§ 264.110-264.120.

89/ 40 C.F.R. §§ 264.140-264.151.

90/ 40 C.F.R. § 264.119.

6.4.11 Interim Status Facilities

All hazardous waste facilities must have a permit to operate.[91/] "Interim status" allows owners and operators of TSD facilities to be treated as having been issued a permit until final administrative action is made on their permit application.[92/] Owners and operators of an existing hazardous waste management facility or a facility in existence on the effective date of RCRA amendments requiring TSD permits are subject to interim status regulation.[93/]

The security, training, waste analysis, emergency, manifesting, recordkeeping, monitoring, closure, and post-closure requirements for TSD facilities seeking or obtaining interim status are similar to those for facilities with final permits.[94/] However, all or part of the groundwater requirements may be waived where an owner or operator can demonstrate that there is a low potential for migration of hazardous wastes or hazardous waste constituents from the facility via the uppermost aquifer to water supply wells or surface water.[95/] An owner or operator that wishes to install a lesser degree of monitoring must have documentation certified by a qualified geologist or geotechnical engineer. The regulations also require the owner or operator of a facility to prepare an outline of a groundwater quality assessment program to be used in the event sampling establishes a suspected discharge from a facility. Upon detecting any suspected discharge, a plan must be submitted to the EPA for assessing groundwater quality. The plan must then be implemented by the owner or operator.

91/ 40 C.F.R. § 270.1.

92/ 40 C.F.R. §§ 264.3, 270.70.

93/ 40 C.F.R. § 270.70.

94/ 40 C.F.R. pt. 265.

95/ 40 C.F.R. §§ 265.90-265.94.

6.4.12 Special Hazardous Wastes and Special Facilities; Universal Wastes

Owners and operators of facilities that treat hazardous waste by chemical, physical, or biological methods other than in tanks, surface impoundments, and land treatment facilities must comply with the standards for interim facilities.96/ Separate rules govern incineration of hazardous waste in boilers and industrial furnaces (BIFs).97/ Special rules also apply to recyclable hazardous waste, recyclable materials utilized for precious metal recovery, and spent lead batteries being reclaimed.98/

In February 1993, EPA proposed a streamlined, reduced regulatory management structure for certain widely generated hazardous wastes known as the "universal wastes" proposal.99/ This proposal, if adopted, would apply to hazardous waste batteries, certain pesticides, and possibly automotive anti-freeze, paint wastes, and mercury-containing items such as light bulbs, thermostats, and thermometers. In July 1994, EPA proposed alternative streamlined regulatory approaches that would apply to mercury-containing lamps.100/

6.5 Used Oil Management

After extensive litigation over whether or not to list used oil as a hazardous waste, in 1992 EPA decided not to list used oil as a hazardous waste and adopted special standards applicable to the management of used oil.101/ Used oil generators must store used oil in containers that are in good condition and not leaking, label containers with the words "Used Oil," respond to releases, ensure that used oil is

---

96/ 40 C.F.R. §§ 265.400-265.406.

97/ 40 C.F.R. §§ 266.100-266.112.

98/ 40 C.F.R. §§ 266.20-266.80.

99/ 58 Fed. Reg. 8102 (Feb. 11, 1993).

100/ 59 Fed. Reg. 38288 (July 27, 1994).

101/ 42 U.S.C. § 6935; 40 C.F.R. pt. 279; 58 Fed. Reg. 26,425 (May 3, 1993); 57 Fed. Reg. 41,612 (Sept. 10, 1992); see also NRDC v. U.S. EPA, 24 ELR 20959 (D.C. Cir. May 27, 1994).

transported off-site by transporters with proper EPA identification numbers, and comply with other requirements set forth in the regulations.[102]/

The use of used oil as a dust supressant is prohibited in Iowa, although a process exists for the DNR to petition approval from EPA for use of used oil as dust supressant.[103]/ Iowa has adopted regulations to govern the collection of waste oil for recycling or reuse.[104]/

6.6 Land Disposal Restrictions

To help prevent groundwater contamination near landfills, the EPA was directed by Congress in the 1984 RCRA Amendments to implement a program prohibiting disposal of hazardous waste in any type of landfill. In 1986, the EPA began phasing out land disposal of all hazardous wastes unless they have been rendered nonhazardous through appropriate treatment.

The EPA divided hazardous wastes into groups which would be prohibited from land disposal in phases. The first group of hazardous waste banned from land disposal as of November 8, 1986, included listed spent solvents (EPA Hazardous Waste Nos. F001-F005)[105]/ and dioxin-containing wastes (F020-F023 and F026-F028).[106]/ As of July 8, 1987, the so-called "California List" wastes were banned which included specified heavy metals, PCBs, and halogenated organic compounds in total concentrations greater than 1,000 mg/kg.[107]/ In August 1988, EPA banned the "first third" wastes which included painting sludges (F006) and many K-listed waste waters and sludges.[108]/ In June 1989, the "second third" wastes including all

---

102/ 40 C.F.R. §§ 279.22, 279.24.

103/ 42 U.S.C. § 6924(l); 40 C.F.R. § 279.82.

104/ Iowa Code § 455B.411(11); Iowa Admin. Code ch. 567-119.

105/ 40 C.F.R. § 268.30.

106/ 40 C.F.R. § 268.31.

107/ 40 C.F.R. § 268.32.

108/ 40 C.F.R. §§ 268.10, 268.33.

F006-F010 wastes and more K-listed and P- and U-listed wastes were banned from land disposal.[109] Finally, in May 1990 (effective August 1990), the "third third" wastes including all remaining listed wastes and all wastes which are hazardous by characteristic were banned.[110] In addition to restricting land disposal, EPA is establishing the recommended treatment method to be used to render wastes nonhazardous and is setting the treatment standard that the waste must meet prior to its disposal.[111] Dilution is prohibited as a substitute for treatment of land-banned hazardous wastes.[112]

The "land ban" or Land Disposal Restriction (LDR) program affects all generators who produce over 100 kilograms (approximately one-half drum) of hazardous waste or one kilogram of acutely hazardous waste per month.[113] These generators must evaluate each waste to determine if it is restricted from land disposal and provide written notification to the off-site treatment, storage, or disposal facility (including recycling and incinerator facilities) of the status of the waste under the LDR program.[114] This notification must accompany the hazardous waste manifest or shipment. In addition, generators are required to keep copies of the notification forms for five years as well as information used in determining whether or not the waste is a land-banned waste. Because the list of LDR waste has been changing annually, and includes many different standards and exceptions, EPA has set up a toll-free hotline to render assistance (800-424-9346).

6.7 RCRA Corrective Action

RCRA corrective action requirements apply to interim status facilities and

---

109/ 40 C.F.R. §§ 268.11, 268.34.

110/ 40 C.F.R. §§ 268.12, 268.35.

111/ EPA's ranking and schedule are published at 40 C.F.R. §§ 268.40-268.45.

112/ 40 C.F.R. § 268.3.

113/ 40 C.F.R. § 268.1(e)(1).

114/ 40 C.F.R. § 268.7.

permitted facilities.[115] The basic steps of corrective action under RCRA are similar to those under the Superfund program discussed in chapter 5 of this handbook. EPA regulations provide that the owner or operator of a facility seeking a TSD permit must institute corrective action as necessary to protect human health and the environment for all releases of hazardous waste or constituents from any solid waste management unit (SWMU) at the TSD facility, regardless of the time at which waste was placed in the SWMU.[116] RCRA corrective action may extend beyond the facility property boundary, and EPA has been expanding the scope of its RCRA corrective action program in recent years. EPA may require corrective action as a condition of its permits by a schedule of compliance when corrective action cannot be completed prior to issuance of the permit. EPA also has authority to respond to imminent and substantial dangers posed by hazardous waste.[117]

6.8 Hazardous Waste Sites and Facilites

Iowa statutes restrict the location of hazardous waste facilities by requiring a proposed owner to obtain a license from the Iowa Environmental Protection Commission (EPC).[118] For these purposes, "facility" does not include land, structures, other appurtenances and improvements contiguous to the source of generation and owned and operated by and exclusively for the treatment, storage, or disposal of hazardous waste of the generator.[119] Application requirements are set forth in Iowa Admin. Code ch. 567-150.

115/ 42 U.S.C. §§ 6924(u)-(v), 6928(h); 40 C.F.R. §§ 264.100-264.101; see also City of Chicago v. Environmental Defense Fund, 114 S. Ct. 1588, 1590 (1994).

116/ 40 C.F.R. § 264.101.

117/ 42 U.S.C. § 6973.

118/ Iowa Code § 455B.443.

119/ Iowa Code § 455B.442(1)(b); see also State v. Presto-X Co., 417 N.W.2d 199 (Iowa 1987) (placing waste in employee's truck was not a "operating a facility for hazardous waste").

6.9 Household Hazardous Waste

Iowa has adopted an number of specific provisions relating to household hazardous materials in Iowa Code ch. 455F and Iowa Admin. Code ch. 567-144 because these wastes are generally exempt from the federal RCRA program. A retailer must obtain a permit to sell household hazardous materials and must provide the consumer with certain information specified in the regulations.[120]

The DNR has established a cleanup program known as Toxic Cleanup Days for household hazardous materials pursuant to Iowa Code § 455F.8. Under this program, collection points throughout the state accept limited amounts of household hazardous materials at a nominal cost. Violations of the household hazardous materials requirements are subject to punishment as a simple misdemeanor.[121]

6.10 Radioactive Waste

Radioactive waste management is controlled by Iowa Code §§ 455B.331-455B.340. The statute defines the term "radioactive material" and directs the DNR to establish policy for the transportation, storage, handling, and disposal of radioactive material.[122] The U.S. Department of Energy is charged with monitoring the federal high-level radioactive waste disposal program under the Nuclear Waste Policy Act.[123]

Iowa statutes provide that a person shall not dispose of, and a sanitary landfill shall not accept for disposal, radioactive materials.[124] Transportation of radioactive materials by highway requires notification of the Iowa Department of

---

120/ Iowa Code § 455F.7; Iowa Admin. Code § 567-144.4.

121/ Iowa Code § 455F.10.

122/ Iowa Code §§ 455B.331, 455B.332.

123/ 42 U.S.C. §§ 10101-10270.

124/ Iowa Code §§ 136C.1, 455B.315.

Transportation and compliance with DOT regulations.[125] The DNR may approve or prohibit the establishment and operation of a nuclear waste disposal site in the state.[126] The DNR must require dispersion modeling of alpha and gamma rays by a person who operates or proposes to operate an incinerator of pathological radioactive materials as well as operating emission limitations.[127]

6.11 Enforcement

6.11.1 Administrative Enforcement Actions

When the EPA discovers violations of pollution laws, it often attempts to resolve the non-compliance through administrative rather than judicial action. Possible administrative resolutions used by the EPA include: (1) negotiation and execution of a stipulation agreement; and (2) issuance of an administrative order.

A stipulation or consent agreement is a written agreement between the EPA and the violator that describes the resolution of the non-compliance situation. This resolution negotiated by the EPA and the violator generally includes: (1) a compliance schedule setting forth specific actions and deadlines to bring the violator into compliance with applicable standards; and (2) a civil penalty for past violations that the violator agrees to pay in exchange for the EPA's waiver of its right to bring a lawsuit against the violator.

An administrative order is a written order issued by the EPA[128] that describes the violations observed by the EPA and contains an order relating to the violations. For example, the EPA may issue an order to revoke or modify a permit or it may issue an order requiring a person to conduct an investigation. An administrative order may assess penalties of up to $25,000 per day of noncompliance.[129]

125/ Iowa Admin. Code § 567-132, 641-39.5.

126/ Iowa Code § 455B.334; Iowa Admin. Code § 567-132.

127/ Iowa Code § 455B.335A.

128/ 42 U.S.C. § 6928(a).

129/ 42 U.S.C. § 6928(a)(3), (c).

6.11.2 Civil Action

Where a negotiated resolution of a non-compliance situation is not possible, or where judicial action is otherwise selected as the appropriate enforcement tool, the EPA may ask the U.S. Department of Justice to commence a civil lawsuit against a person who has failed to comply with applicable standards. A civil lawsuit may seek any one or more of the following:

1. Recovery of civil penalties in an amount of up to $25,000 per violation per day;

2. Imposition of an injunction against the violator of a rule, standard, order, stipulation agreement, or permit;

3. Issuance of an order requiring the violator to undertake some action;

4. Forfeiture or payments of a sum which will adequately compensate the EPA for the reasonable value of cleanup and other expenses directly resulting from an unauthorized discharge; and

5. Other appropriate relief.

If the EPA finally prevails in a civil action, the EPA may also recover the reasonable value of all or part of its litigation expenses.[130]

6.11.3 Criminal Action

RCRA violations may be punished as crimes by a fine of not more than $50,000 for each day of violation or imprisonment not to exceed two years or both.[131] RCRA violations that involve knowing endangerment of death or serious bodily injury may be punished as a crime by fine of not more than $250,000 for individuals ($1,000,000 for organizations) and 15 years imprisonment or both.[132]

130/ 42 U.S.C. §§ 6928, 6972, 6973.

131/ 42 U.S.C. § 6928(d).

132/ 42 U.S.C. § 6928(e).

Although RCRA refers to "knowing" violations, courts have not required that a defendant "know" the waste was "hazardous waste" or that he or she was violating the law. Instead, the prosecutor must show only that the defendant knew that he or she was disposing of waste which had the potential to be harmful to convict a defendant of RCRA violations.133/

6.11.4 Citizen Suits

Subject to notice and timing restrictions, RCRA permits citizen suits to respond to regulatory violations or to an "imminent and substantial endangerment" as long as EPA has not commenced an enforcement action.134/ In most circumstances, the notice requirement is considered jurisdictional.135/ Remedies include injunctive relief compelling compliance, restraining actions constituting an imminent hazard, and ordering "such other action as may be necessary" as well as awards of civil penalties for hazardous waste violations, attorneys fees, and expenses136/ Under Iowa's citizen suit statute, any person adversely affected may commence a civil action against any person for violating Iowa's hazardous waste statutes and rules.137/

6.11.5 Iowa Enforcement Authority

While EPA administers the hazardous waste management program in Iowa, Iowa statutes relating to hazardous waste enforcement remain in place. Under these

---

133/ E.g., United States v. Laughlin, 10 F.3d 961, 966 (2d Cir. 1993); United States v. Hoflin, 880 F.2d 1033, 1039 (9th Cir. 1989), cert. denied, 493 U.S. 1083 (1990); United States v. Hayes Int'l Corp., 786 F.2d 1499, 1505 (11th Cir. 1986) (lack of knowledge that waste was hazardous or that permit was required held invalid as defenses).

134/ 42 U.S.C. §§ 6972(a)(1)(B) and 6972(b)(2)(B); Coburn v. Sun Chem. Corp., 19 Envtl. L. Rep. 20256, 20260-61 (E.D. Pa. 1988).

135/ 42 U.S.C. § 6972(b)(1)(A); Hallstrom v. Tillamook, 493 U.S. 20 (1989).

136/ 42 U.S.C. § 6972(a); see also Lincoln Prop. Ltd. v. Higgins, 36 Env't Rep. Cas. (BNA) 1228, 1242-43 (E.D. Cal. Jan. 21, 1993) (issuing injunctive relief under RCRA).

137/ Iowa Code § 455B.111.

statutes, the DNR has authority to issue administrative enforcement orders and may ask that the attorney general institute legal proceedings to obtain compliance with administrative orders, seek injunctive relief, or seek penalties.[138/] Violations of hazardous waste management requirements may be subject to a civil penalty of not to exceed $10,000 for each day of violation.[139/]

Criminal violations of hazardous waste management requirements may be punished by $25,000 fine or one year imprisonment or both for each day of violation.[140/] In certain circumstances, these criminal penalties may be higher.[141/]

138/ Iowa Code §§ 455B.388, 455B.418.

139/ Iowa Code § 455B.417(3); see also Iowa Code §§ 455B.430, 455B.454, 455B.466.

140/ Iowa Code § 455B.417(2).

141/ Iowa Code § 455B.417(2) (subsequent convictions), ch. 716B (knowing violations); see also State v. Hunt, 512 N.W.2d 285, 290 (Iowa 1994) (upholding criminal convictions under § 716B.2 for disposal at unauthorized location and under § 716B.4 for storage without a permit).

CHAPTER SEVEN

INFECTIOUS WASTE MANAGEMENT

7.0 Agencies Regulating Infectious Waste

A number of federal and state agencies regulate infectious waste, often with a significant amount of overlap. The United States Environmental Protection Agency (EPA) has issued detailed guidelines for infectious waste management.[1] The Occupational Safety and Health Administration (OSHA) also regulates infectious waste in the workplace.[2] In addition, the Centers for Disease Control (CDC) provides guidance and information regarding infectious waste.[3]

The Iowa Department of Natural Resources (DNR) has general regulatory authority over generators of infectious waste such as hospitals and clinics, incinerators, landfills, and transporters under existing statutes and rules, and it is responsible for enforcement involving infectious waste.[4] The DNR administers the infectious waste management program in cooperation with the Iowa Department of Public Health. Finally, the Iowa Department of Employment Services Division of Labor regulates infectious waste in the workplace.[5]

7.1 Iowa Infectious Waste Management Act

The Iowa Infectious Waste Control Act[6] regulates the management of infectious waste in the state. It governs persons and entities that generate, handle, transport, or dispose of infectious, pathological, and related wastes. The purposes of the Act include preventing the spread of disease, minimizing opportunities for

---

1/ 40 C.F.R. §§ 259.30-259.91.

2/ 29 C.F.R. § 1910; Executive Order 12196.

3/ See, e.g., Morbidity and Mortality Weekly Reports, Aug. 21, 1987 and June 24, 1988.

4/ Iowa Code § 455B.501(2).

5/ Iowa Code ch. 88 and 89B; Iowa Admin. Code ch. 347-110.

6/ Iowa Code §§ 455B.501-455B.505.

accidental contact with infectious agents, and ensuring proper disposal of such wastes.

7.1.1 Definitions

Just as federal and state agencies have overlapping jurisdiction over infectious waste, the various agencies have promulgated their own definitions.7/ The Iowa Infectious Waste Management Act8/ defines "infectious waste" and specific types of such waste as follows:

> a. "Infectious" means containing pathogens with sufficient virulence and quantity so that exposure to an infectious agent by a susceptible host could result in an infectious disease when the infectious agent is improperly treated, stored, transplanted, or disposed.
>
> b. "Infectious waste" means waste, which is infectious, including but not limited to contaminated sharps, cultures, and stocks of infectious agents, blood and blood products, pathological waste, and contaminated animal carcasses from hospitals or research laboratories.
>
> c. "Contaminated sharps" means all discarded sharp items derived from patient care in medical, research, or industrial facilities including glass, vials containing materials defined as infectious, hypodermic needles, scalpel blades, and pasteur pipettes.
>
> d. "Cultures and stocks of infectious agents" means specimen cultures collected from medical and pathological laboratories, cultures and stocks of infectious agents from research and industrial laboratories, wastes from the production of biological agents, discarded live and attenuated vaccines, and culture dishes and devices used to transfer, inoculate, or mix cultures.

7/ For example, EPA includes infectious waste in its definition of "hazardous waste" in the Resource Conservation and Recovery Act of 1976 (RCRA), 42 U.S.C. § 6903(5). In addition, EPA's guide for infectious waste management, published in May 1986, states a broad definition of the term. RCRA also uses the term "medical waste." 42 U.S.C. § 6903(40). EPA's medical waste tracking pilot program applicable in five eastern states uses the term "regulated medical waste." 40 C.F.R. § 259.30.

8/ Iowa Code § 455B.501(1).

e. "Human blood and blood products" means human serum, plasma, other blood components, bulk blood, or containerized blood in quantities greater than twenty milliliters.

f. "Pathological waste" means human tissues and body parts that are removed during surgery or autopsy.

g. "Contaminated animal carcasses" means waste including carcasses, body parts, and bedding of animals that were exposed to infectious agents during research, production of biologicals, or testing of pharmaceutical.

Iowa's hazardous waste and substance management laws also define "hazardous waste" to include certain wastes or combination of wastes with "infectious characteristics."9/

7.2 Waste Segregation, Labeling, and Handling

Under DNR policy, untreated infectious waste should be segregated from other waste material at its point of generation and maintained in separate packaging throughout its collection, storage, and transportation. Infectious waste should be packaged and transported in a manner that prevents release of the waste material. Bags, boxes, and other containers used to collect, transport, or store infectious waste should be clearly labeled with a biohazard symbol or with the words "infectious waste" written in letters no less than one inch in height. Containers which have been in direct contact with infectious waste must be disinfected prior to reuse.

Sharps should be placed in puncture-resistant containers. They should not be compacted or mixed with other waste material and should not be disposed of at a refuse-derived fuel facility or any facility where waste is hand-sorted.

Further guidance is available in the EPA regulations adopted for the medical waste tracking pilot program applicable in Connecticut, New Jersey, New York, Rhode Island, and Puerto Rico pursuant to the Medical Waste Tracking Act of

---

9/ Iowa Code § 455B.411.

1988.[10/] EPA is evaluating the expansion of this pilot program.[11/] In addition, EPA is in the process of drafting regulations intended to reduce air emissions from medical waste incinerators and include operator training and certification requirements that are expected to be finalized in fall 1995. EPA expects the proposed regulations to increase the cost of incinerating medical waste.

7.3 Treatment and Disposal Alternatives and Restrictions

EPA's *Guide for Infectious Waste Management* recommends different treatment techniques for different types of infectious waste, with alternatives available for nearly each type of waste:

1. Steam sterilization (autoclaving) is recommended for isolation wastes, cultures and stocks of infectious agents and associated biologicals; human blood and blood products; pathological wastes; contaminated sharps; and contaminated animal carcasses and body parts. For pathological wastes and contaminated animal carcasses and body parts, EPA recommends that, for aesthetic reasons, steam sterilization should be followed by incineration or by grinding and then flushing to the sewer system.

2. Incineration is recommended for each of the wastes for which steam sterilization is an alternative, and is also recommended for contaminated animal bedding.

3. Thermal inactivation is recommended for cultures and stocks of infectious agents and associated biologicals.

4. Chemical disinfection is recommended for liquid forms of cultures and stocks of infectious agents and associated biologicals and human blood and blood products.

10/ 42 U.S.C. §§ 6992-6992k; 40 C.F.R. §§ 259.40-259.45; see also 49 C.F.R. § 171.14(b)(6) and 59 Fed. Reg. 53,116 (Oct. 21, 1994) (regarding DOT medical waste transportation policy).

11/ See generally Note, Infectious Waste: A Guide to State Regulation and a Cry for Federal Intervention, 66 Notre Dame L. Rev. 555 (1990).

5. Discharge to sanitary sewer for treatment in municipal sewage system is recommended for human blood and blood products (provided that secondary treatment is available).

6. Handling by a mortician (burial or cremation) is recommended for pathological wastes.

Approximately thirty hospitals in Iowa are currently authorized by the DNR to incinerate their infectious waste, and two hospitals have authority to implement chemical treatment of their infectious waste.

The Iowa legislature has directed the DNR and the Environmental Protection Commission to adopt rules which require a person who owns or operates an infectious medical waste collection or transportation operation to obtain an operating permit prior to initial operation.12/ The rules are expected to address the areas of operator safety, recordkeeping and tracking procedures, best available appropriate technologies, emergency response and remedial action procedures, waste minimization procedures, and long-term liability. At the time of this writing, the rules had not yet been drafted.

For state universities that operate an infectious medical waste incinerator, the DNR and Department of Public Health are to require periodic monitoring to measure compliance with emission limitations standards for toxic air pollutants.13/ If monitoring results do not meet the emission limitations standards established, the DNR shall require that the university employe the best available control technology (BACT) for toxics. Following employment of BACT, if the incinerator does not meet the standards established, the permit for operation of the infectious medical waste incinerator will be revoked.

12/ Iowa Code § 455B.504.

13/ Iowa Code § 455B.502.

7.4 Infectious Waste Management Plan

Although not currently mandated by Iowa law, many facilities that generate, store, decontaminate, incinerate, or dispose of infectious waste prepare a management plan for infectious or pathological waste handled by the facility that describes, to the extent applicable, the following:

1. The type of infectious waste and pathological waste generated or handled;

2. The segregation, packaging, labeling, collection, storage, and transportation procedures for the infectious or pathological waste that will be followed;

3. The decontamination or disposal methods for the infectious or pathological waste that will be used;

4. The transporters and disposal facilities that will be used;

5. The steps that will be taken to minimize the exposure of employees to infectious agents throughout the process of disposing of infectious or pathological wastes; and

6. The name of the individual responsible for the management of the infectious or pathological waste.

7.5 Moratorium on Infectious Waste Treatment and Disposal Facilities

The Iowa legislature established a number of requirements applicable to new commercial infectious waste treatment and disposal facilities. For example, the DNR is prohibited from granting a permit for the construction or operation of a commercial infectious waste treatment or disposal facility within one mile of a site or building which has been placed on the national register of historic places.14/ This restriction, however, does not apply to hospitals, certain licensed health care facilities, physicians' offices or clinics, and other health service-related entities.

14/ Iowa Code § 455B.505.

More significantly, the DNR will not grant permits for the construction or operation of a commercial infectious waste treatment or disposal facility until the Environmental Protection Commission has adopted rules addressing operator safety, recordkeeping and tracking procedures, best available appropriate technologies, emergency response and remedial action procedures, waste minimization procedures, and long-term liability.15/ The DNR also is to adopt rules that require a person who owns or operates an infectious waste treatment or disposal facility to obtain an operating permit before initial operation of the facility. Given that the rules have not yet been proposed or adopted, Iowa Code § 455B.503 arguably prohibits the construction or initial operation of a commercial infectious waste treatment or disposal facility.

Prior to the siting of a proposed, new sanitary landfill, incinerator, or infectious medical waste incinerator, a city, county, or private agency must submit a request for local siting approval to the city council or county board of supervisors which governs the city or county in which the proposed site is to be located.16/ The city council or county board of supervisors must approve or disapprove the site. Approval may be granted only if the proposed project satisfies a number of specific criteria set forth in the statute. In addition, the DNR must be consulted with and a notice and public hearing process must be followed as part of the approval.17/

7.6 Enforcement

As noted above, infectious waste is in some circumstances regarded as a "hazardous waste" for purposes of Iowa Code § 455B.411. As a result, the hazardous waste disposal penalties in Iowa Code chapter 716B arguably would be applicable to transportation, storage, treatment, or disposal violations involving infectious waste.

15/ Iowa Code § 455B.503.

16/ Iowa Code § 455B.305A.

17/ Iowa Code § 455B.305A.

CHAPTER EIGHT

RELEASE REPORTING AND RIGHT-TO-KNOW

8.0 Introduction To Iowa Reporting Requirements

This chapter discusses requirements as to when a release of hazardous substances must be reported to relevant agencies, laws governing the public's and employees' right to know about hazardous substance uses that may affect them. Under Iowa Code § 455B.386, a person manufacturing, storing, handling, transporting, or disposing of a hazardous substance must notify the DNR and the local police department or the office of the sheriff of the affected county of the occurrence of a hazardous condition as soon as possible but not later than six hours after the onset of the hazardous condition or discovery of the hazardous condition at:

Iowa DNR Spill Reporting -- 515-281-8694

8.1 Federal Hazardous Substance Release Reporting Requirements

8.1.1 Comprehensive Environmental Response, Compensation and Liability Act (CERCLA) Reportable Quantities

The federal Comprehensive Environmental Response, Compensation and Liability Act (CERCLA or Superfund)[1] is better known for its liability provisions discussed in chapter 5 of this handbook relating to cleaning up contaminated sites. CERCLA, however, also establishes release reporting requirements in section 103. Any person in charge of a facility must, "as soon as he has knowledge of any release"[2] of a reportable quantity of a hazardous substance, immediately notify:

National Response Center -- 800-424-8802

1/ 42 U.S.C. §§ 9601-9675.

2/ 42 U.S.C. § 9603.

A written follow-up report should be provided to the National Response Center and the DNR. Reportable quantities (RQ) of CERCLA-listed hazardous substances are set forth in Table 302.4 following 40 C.F.R. § 302.4. RQs for different substances vary widely. For example, the RQ for polychlorinated biphenyls (PCBs) is 1 pound, while the RQ for hydrochloric acid is 5,000 pounds. To prepare for an emergency spill, a facility that handles hazardous substances should maintain a current list of RQs for each hazardous substance that could be released. For "continuous releases" of hazardous substances above RQs, which takes place at approximate 10,000 facilities in the United States, EPA has adopted a special reporting form.3/

8.1.2 Emergency Planning and Community Right-to-Know (EPCRA) or SARA Title III

In 1986, the U.S. Congress adopted the Superfund Amendments and Reauthorization Act (SARA), which amended CERCLA. SARA Title III, which was established largely in response to the Bhopal chemical release, is commonly known as the Emergency Planning and Community Right-to-Know Act (EPCRA), 42 U.S.C. §§ 11001-11050. EPCRA established a number of requirements applying to each company that manufactures, stores, or uses chemicals that are designed to assist local agencies in emergency planning for an accidental release, fire, or other disaster. As discussed below, EPCRA § 313 requires owners or operators of certain facilities that manufacture or otherwise use toxic chemicals to report annually their releases of these chemicals to each environmental medium. In addition, EPCRA obligates states to establish emergency response commissions and local emergency planning committees that are charged with developing and implementing plans to respond to emergency releases of hazardous chemicals.

3/ 40 C.F.R. §§ 302.8, 355.40. Note that reportable quantities of the same hazardous substance may differ under different statutes. Compare 40 C.F.R. § 302.4 (CERCLA) with 40 C.F.R. pt. 355, App. A (EPCRA) and 40 C.F.R. § 117.3 (Clean Water Act).

Emergency planning requirements under EPCRA are triggered if a facility has an extremely hazardous substance on site in an amount equal to or exceeding its threshold planning quantity (TPQ). The list of extremely hazardous substances and their TPQs and RQs is set forth in 40 C.F.R. Part 355, App. A. Paralleling the CERCLA requirements, EPCRA requires a facility owner or operator to notify the state of any release of an extremely hazardous substance above RQs.

8.1.3 Clean Water Act (CWA) Discharge Reporting

The federal Clean Water Act contains requirements for spill reporting and the development of emergency prevention and response plans. Under Section 311 of the Clean Water Act,[4] any person in charge of a vessel or facility must immediately notify the National Response Center as soon as he or she has knowledge of any discharge of oil or a hazardous substance above the reportable quantities set forth in 40 C.F.R. § 117.1.[5]

8.1.4 Clean Air Act Reporting and Risk Management Plans

The Clean Air Act Amendments of 1990 added a new provision that focuses on prevention of accidental releases of extremely hazardous chemicals.[6] Under this provision, owners and operators of facilities that handle toxic chemicals listed under EPCRA Section 313 and certain other chemicals must identify hazards that may result from accidental releases of these chemicals.

EPA has proposed regulations requiring these facilities to prepare and implement risk management plans designed to prevent accidental releases and establish procedures for prompt emergency response if a release occurs. When the

---

4/ 33 U.S.C. § 1321(b)(5).

5/ 33 C.F.R. § 153.203.

6/ 42 U.S.C. § 7412(r); 58 Fed. Reg. 54,190 (Oct. 20, 1993) (proposing implementing regulations); 59 Fed. Reg. 4478 (Jan. 31, 1994) (listing threshold quantities for accidental release prevention). See Delhotal, The General Duty to Prevent Accidental Releases of Extremely Hazardous Substances, 13 J. Energy Nat. R. & Envtl. L. 61 (1993).

regulations take effect, the plan must include a five-year history of past releases, an evaluation of the worst-case scenario of a potential release, and specific programs for preventing and responding to accidental releases. EPA established threshold planning requirements in regulations published in January 1994.7/

8.2 Hazardous Waste Release Response And Reporting

In the event of a fire, explosion, or other release of hazardous waste which could threaten human health outside the facility or when a hazardous waste generator has knowledge that a spill has reached surface water, the generator must immediately notify the U.S. Coast Guard National Response Center. The report should include the name of the company or individual responsible, the address, the material involved, the quantity, the time and date of incident, and the cause of the incident if known.

In addition to these notification requirements, a person who generates hazardous waste that spills is required to recover it as soon as is practicable and to take such other action as may be reasonably necessary to protect health and the environment.8/ For any fires, explosions, or other releases of hazardous waste or hazardous waste constituents to land, air, or water, a hazardous waste treatment, storage, or disposal (TSD) facility is responsible for making arrangements with local authorities to prepare for an emergency.9/ At least one employee on the facility premises or on call must be responsible for coordinating emergency response measures in the event of a release at the facility, including notifying federal, state, and local agencies.10/

7/ 59 Fed. Reg. 4478 (Jan. 31, 1994).

8/ 40 C.F.R. §§ 262.34 (d)(5) (generator requirements), 263.30-263.31 (discharges during transportation), 264.56(g) (emergency procedures).

9/ 40 C.F.R. §§ 264.50-264.53.

10/ 40 C.F.R. § 264.55; see also section 6.2.9 of this handbook.

8.3 Iowa "Hazardous Condition" Notification Laws

As noted above, Iowa Code § 455B.386 provides that a person manufacturing, storing, handling, transporting, or disposing of a hazardous substance must notify the DNR and the local police department or the office of the sheriff of the affected county of the occurrence of a hazardous condition as soon as possible but not later than six hours after the onset of the hazardous condition or discovery of the hazardous condition. This statute applies the "hazardous condition" definition discussed in chapter 5 of this handbook.

Iowa Administrative Code ch. 567-131 prescribes in more detail the notification requirements, including requirements relating to notification of the Department of Public Defense Disaster Services Division (515-281-3231). Verbal reports should be followed by a written report within 30 days. All subsequent finding and laboratory results should be reported and submitted in writing to the DNR as soon as they become available.11/

According to DNR reporting guidelines, a hazardous condition exists, if, without intervention:

1. The hazardous substance has the potential to leave the property via overland runoff, or through a drain, tile line, culvert, or other conduit.

2. The hazardous substance has the potential to reach a water of the state--either surface water or groundwater.

3. The hazardous substance can be detected in the air at the boundaries of the property by either senses or monitoring equipment.

4. People other than the responsible party's employees can be potentially exposed to the hazardous substance.

5. Local officials respond to the incident.

6. The amount of hazardous substance released exceeds the federal reportable quantity (RQ).

11/ Iowa Admin. Code § 131.1(3).

8.4 Iowa Underground Storage Tank Release Notification Requirements

As discussed in more detail in chapter 10 of this handbook, under the Iowa regulations relating to underground storage tanks (USTs), owners and operators of UST systems must submit reports of all releases including suspected releases, spills and overfills, and confirmed releases.12/ Owners and operators of UST systems must report to the DNR within 24 hours (or within 6 hours if a hazardous condition exists) for any spill or overfill of petroleum that results in a release to the environment that exceeds 25 gallons, exceeds a CERCLA RQ, or that causes a sheen on nearby surface water.13/

With regard to suspected releases, owners and operators of UST systems must report: (a) the discovery of released regulated substances at the site or in the surrounding area (such as the presence of free product or vapors in soils, basements, sewer and utility lines, or nearby surface water); (b) unusual operating conditions (such as erratic behavior of product dispensing equipment, sudden loss of product, or unexplained presence of water in the tank), unless system equipment is found to be defective but not leaking and is immediately repaired or replaced; and (c) monitoring results from a release detection method that indicate a release may have occurred unless the monitoring device is found to be defective or, in the case of inventory control, a second month of data does not confirm the initial result.14/ Confirmed releases must be reported within 24 hours or within another reasonable period of time determined by the DNR.15/

8.5 Oil Spills and the Oil Pollution Act (OPA)

Following the Exxon Valdez oil spill in Alaska, Congress adopted the Oil

---

12/ Iowa Admin. Code § 135.4(5)(a)(2).

13/ Iowa Admin. Code § 135.6(4)(a). As noted above, spills exceeding a CERCLA RQ must be reported immediately (rather than within 24 hours).

14/ Iowa Admin. Code § 135.6(1).

15/ Iowa Code § 135.7(2).

Pollution Act (OPA)16/ to supplement the CWA provisions relating to oil spills. With regard to spill notification, OPA added provisions codified within the CWA at 33 U.S.C. § 1321(j) establishing a National Response Unit within the Coast Guard with responsibility for administering Coast Guard responses to spills of oil or hazardous substances in navigable waters.

For offshore vessels and facilities as well as onshore facilities that could reasonably be expected to discharge into navigable waters, OPA provides that response plans must be prepared that, among other things, require immediate notification of the Coast Guard. EPA regulations requiring onshore facility response plans apply to: (1) facilities that have a total oil storage capacity greater than 42,000 gallons which transfer oil over water to or from vessels; and (2) certain other facilities with oil storage capacity greater than 1 million gallons.17/

8.6 Toxic Substances Control Act (TSCA) and PCB Spills

Reporting requirements under the Toxic Substances Control Act (TSCA)18/ focus primarily on persons who manufacture, process, or distribute chemical substances in commerce. EPA regulations issued under TSCA, however, are the primary federal authority governing polychlorinated biphenyls (PCBs). For PCB spills that take place after May 4, 1987, all spills involving 10 pounds or more of PCBs must be reported to the National Response Center and the EPA Regional Office of Pesticides and Toxic Substances Branch in the shortest possible time after discovery, but in no case later than 24 hours after discovery.19/ In addition, all spills of PCBs above 50 parts per million (ppm) that directly contaminate surface waters,

---

16/ 33 U.S.C. §§ 2701-2761; see also 59 Fed. Reg. 34,070 (July 1, 1994) (promulgating regulations).

17/ 59 Fed. Reg. 34,070 (July 1, 1994) (to be codified as part of the SPCC regulations at 40 C.F.R. § 112.20); see also chapter 10 of this handbook regarding storage tank requirements.

18/ 15 U.S.C. §§ 2601-2671

19/ 40 C.F.R. § 761.125(a)(1).

sewers, drinking water supplies, grazing lands, or gardens must be reported immediately to the EPA Regional Office of Pesticides and Toxic Substances Branch.

8.7 Transportation Incidents

8.7.1 Federal Hazardous Materials Transportation Act (HMTA) Notifications

The federal Hazardous Materials Transportation Act (HMTA), 49 U.S.C. §§ 5101-5127, is administered by the U.S. Department of Transportation (DOT). HMTA regulations, sometimes called "Hazmat regulations," govern the transportation of hazardous materials identified in 49 C.F.R. § 172.101 by air, highway, rail, or water. HMTA regulations require that, at the earliest practicable moment, each carrier who transports hazardous materials must notify the DOT (800-424-8802) of any incident that occurs during the course of transportation in which death or injury occurs, property damage exceeds $50,000, or a release of radioactive material or an etiologic agent occurs.20/ In addition, a written report must be filed on DOT Form 5800.1 of any such incident or when there has been an unintentional release of hazardous materials from a package (including a tank) or any quantity of hazardous waste has been discharged during transportation.21/ Additional requirements apply to transportation by pipeline.22/

8.7.2 Iowa Transportation Incident Notification Requirements

Iowa law expressly incorporates certain HMTA requirements relating to transportation of hazardous materials, including the HMTA definition of hazardous materials.23/ In the case of an accident involving the transportation of a hazardous

---

20/ 49 C.F.R. § 171.15(a).

21/ 49 C.F.R. § 171.16.

22/ See, e.g., 49 U.S.C. § 2000; Iowa Code ch. 479; see also Kinley Corp. v. Iowa Util. Bd., 999 F.2d 354, 359 (8th Cir. 1993) (federal hazardous liquid pipeline act preempts Iowa Code ch. 479 with regard to jet fuel pipeline); ANR Pipeline Co. v. Iowa State Commerce Comm'n, 828 F.2d 465 (federal act preempts Iowa Code ch. 479 regarding natural gas pipeline).

23/ Iowa Code § 321.1(31).

material, a carrier transporting hazardous material upon a public highway must notify the police or a peace officer in the county or city in which the accident occurs.[24] In most circumstances, hazardous condition notification requirements discussed in section 8.3 would apply to such incidents. A person who violates this notification requirement is guilty of a serious misdemeanor.

8.8 Occupational Safety and Health Act (OSHA) Hazard Communication Standard and Process Safety Management Regulations

The federal Occupational Safety and Health Act (OSHA)[25] contains numerous provisions relating to emergency response training and emergency plans. In particular, the OSHA Hazard Communication Standard, which applies to all facilities with employees who may be exposed to hazardous chemicals, mandates development of a hazard communication program that includes employee training requirements and emergency response procedures.[26] Special emergency response training for hazardous waste operations and emergency response (HAZWOPER) employees is required.[27]

In addition to the OSHA Hazard Communication Standard requirements discussed in more detail in the following section of this handbook, OSHA regulations were adopted in 1992 requiring process safety management of highly hazardous chemicals.[28] The process safety management regulations apply to "highly hazardous chemicals" listed in the regulation and certain flammable liquids. Their purpose is to prevent or minimize the consequences of catastrophic releases that present a serious danger to employees in the workplace. These process safety management regulations require employee training, process hazard analysis,

---

24/ Iowa Code § 321.266; see also Iowa Admin. Code § 761-640.3.

25/ 29 U.S.C. § 655(b).

26/ 29 C.F.R. §§ 1910.120, 1910.1200.

27/ 29 C.F.R. § 1910.120.

28/ 29 C.F.R. § 1910.119; see also section 17.3 of this handbook.

emergency planning and response, incident investigation, and compliance auditing. Employee emergency action planning and training requirements incorporated by these regulations include fire and emergency reporting obligations.29/

8.9 Employee Right-To-Know

8.9.1 Federal Employee Right-To-Know

OSHA has estimated that over 32 million workers are exposed to approximately 650,000 hazardous chemical products in 3.5 million workplaces in the United States.30/ To provide employees with information about the hazards and identities of chemicals in the workplace and the measures employees can take to protect themselves, the federal government has adopted a number of laws. The federal government's efforts have resulted in regulation primarily through four statutes: OSHA31/ which is administered by the Occupational Safety and Health Administration; the Toxic Substances Control Act of 1976 (TSCA)32/ which is administered by the EPA; the National Labor Relations Act (NLRA)33/ which is administered by the National Labor Relations Board (NLRB); and EPCRA,34/ which is administered by EPA as well.

OSHA outlines the underlying concerns for safety in the workplace.35/ The Act sets maximum exposure standards for specific substances, and it requires that warnings be given to employees exposed to hazardous substances for which an exposure standard has been set.36/ OSHA's Hazard Communication Standards

---

29/ 29 C.F.R. §§ 1910.38(a), 1910.119(n).

30/ 59 Fed. Reg. 6126 (Feb. 9, 1994).

31/ 29 U.S.C. §§ 651-678.

32/ 15 U.S.C. §§ 2601-2671.

33/ 29 U.S.C. §§ 151-169.

34/ 42 U.S.C. §§ 11001-11050. See sections 8.1.2 and 8.10.1 of this handbook.

35/ 29 U.S.C. § 655(b)(5).

36/ 29 U.S.C. § 655(b)(7).

create requirements regarding availability of information to workers about toxic and hazardous substances in the workplace.37/ Following litigation regarding whether the Hazard Communication Standards should apply in the nonmanufacturing sector, in February 1994 OSHA adopted amended regulations applicable uniformly to manufacturing and nonmanufacturing, shipyard, marine, longshoring, construction, and agricultural employment.38/

Material Safety Data Sheets (MSDSs) are the centerpiece of the Hazard Communication Standards. The MSDS usually states the trade and chemical name of the substance, the hazardous components, physical characteristics, data concerning fire and explosion hazards, health hazard information, stability or reactivity of the substance, procedures in case of spillage or leakage, safety and handling cautions, and other personnel protection data. Chemical manufacturers and importers must obtain or develop an MSDS for each hazardous chemical they produce or import.39/

Employers are required to assemble a list of hazardous materials in the workplace, label all chemicals, provide employees with access to MSDSs, and provide education and training.40/ Containers must be labeled with the identity of the contents, appropriate hazard warnings, and (unless transferred to another container after purchase) the name and address of the manufacturer.41/ Employee training programs are required to inform employees of the labels and MSDSs and to make them aware of acts that are required to avoid or minimize exposure to

---

37/ 29 C.F.R. § 1910.1200.

38/ 59 Fed. Reg. 6126 (Feb. 9, 1994) (to be codified at 29 C.F.R. §§ 1910.1200, 1915.1200, 1917.28, 1918.90, and 1926.59).

39/ 29 C.F.R. § 1910.1200(g)(1).

40/ 29 C.F.R. § 1910.1200(g).

41/ 29 C.F.R. § 1910.1200(f).

hazardous chemicals.42/ Employee training must be provided at the time of initial assignment or whenever a new hazard is introduced into their work area.43/

8.9.2 Employees and Iowa Hazardous Chemicals Risks Right-To-Know

The Iowa Occupational Safety and Health Act (IOSH or IOSHA), Iowa Code chapter 88, is most well-known for its regulations governing workplace exposure to hazardous conditions. IOSH requires employers to notify employees promptly of exposure to toxic materials or harmful physical agents in concentrations above permissible exposure limits or other IOSH standards.44/

Iowa's Hazardous Chemicals Risks Right-To-Know Act standards are found in Iowa Code chapter 89B and in the rules of the state's Department of Employement Services Division of Labor.45/ The rules apply to "hazardous chemicals," meaning any chemical that is a physical hazard or a health hazard, which are in turn broadly defined in the rules.46/

The Act provides that employees have the right to be informed about the hazardous chemicals to which the employee may be exposed in the workplace, the potential health hazards of the hazardous chemicals, and the proper handling techniques for the hazardous chemicals.47/ An employer must provide or make this information available to an employee.48/ The Iowa legislature directed that the Iowa rules be based on the federal OSHA hazard communication regulations at 29 C.F.R. Part 1910.1200. In Iowa Admin. Code chapter 347-10, the Iowa Department of

---

42/ 29 C.F.R. § 1910.1200(h).

43/ 29 C.F.R. § 1910.1200(h).

44/ Iowa Code § 88.5(2) & (4); see also Iowa Admin. Code ch. 347-2.

45/ Iowa Admin. Code ch. 347-110 to 140, promulgated pursuant to Iowa Code ch. 88 and 89B. See generally Note, Hazardous Chemicals -- Right to Know in Iowa, 36 Drake L. Rev. 419 (1986-87).

46/ Iowa Admin. Code § 347-110.2.

47/ Iowa Code § 89B.8.

48/ Iowa Code § 89B.8; Iowa Admin. Code § 347-120.6.

Employment Services Division of Labor adopted the federal standards of 29 C.F.R. Part 1910 by reference.

Employers are required to develop and implement a written Employee Right-To-Know program which, at a minimum, describes how the training, availability of information, and labeling provisions of the rules will be met. The rules prescribe the content of the written program and the frequency of training.49/ Availability of written information for employees, including MSDS requirements and hazardous substance labeling requirements, are also prescribed.50/

8.10 Community Right-To-Know

8.10.1 Federal Community Right-To-Know

As noted above, EPCRA requires that each company that manufactures, stores, or uses chemicals reports its inventories to local agencies to assist in emergency planning regarding these hazardous substances in the event of an unplanned release, fire, or similar disaster. EPCRA required the creation of state emergency response commissions and local emergency planning committees which were charged with developing and implementing plans to respond to emergency releases.51/

EPCRA requires three different sets of reports concerning chemical substances. The first two concern inventory-related data on "hazardous chemicals," which must be submitted to the appropriate local emergency planning committe, the state emergency response commission, and to the local fire department.52/ In general terms, an owner or operator of a facility that is required to have available an MSDS must submit the MSDS or a list of the hazardous chemicals and an

---

49/ Iowa Admin. Code §§ 347-120.3, 120.6.

50/ Iowa Admin. Code §§ 347-120.4, 120.5.

51/ 42 U.S.C. § 11001.

52/ 42 U.S.C. §§ 11021(a)(1), 11022(a)(1).

emergency and hazardous chemical inventory form.[53] These forms are commonly referred to as the Tier 1 and Tier 2 forms.

Minimum thresholds for Tier 1 and Tier 2 reporting are: (1) 500 pounds or the threshold planning quantity (TPQ) for extremely hazardous substances designated under EPCRA § 302, 42 U.S.C. § 11,002; or (2) 10,000 pounds at any one time during the preceding calendar year for all other hazardous chemicals for which a facility is required to have an MSDS.[54] Because the Tier 2 form covers the requirements for both Tier 1 and Tier 2, most facilities simply file the Tier 2 forms. Petroleum products (above 10,000 pounds) are one of the most common substances triggering an obligation to file a Tier 2 form.

The third set of EPCRA reports is the more well-known annual toxic release inventory (TRI) reports submitted on EPA Form R pursuant to EPCRA § 313. Section 313 applies to facilities that have 10 or more full-time employees, are listed in Standard Industrial Classification (SIC) Codes 20 through 39, and manufacture, process, or otherwise use toxic chemicals on the "Title III List of Lists." At the time of this writing, the EPA is considering the addition of approximately 300 more toxic chemicals to this list, including more than 150 pesticides. The threshold level for reporting is 10,000 pounds in most circumstances. Section 313 obligates owners and operators of these facilities to report annually to EPA and the DNR the quantities of those substances present at the facility, how those chemicals are treated or disposed of, and the annual quantity of each toxic chemical entering each environmental medium.[55] EPCRA creates a number of civil, criminal, and administrative penalties for violations of statutory reporting requirements[56] and authorizes

---

53/ 40 C.F.R. §§ 370.21, 370.25.

54/ 42 U.S.C. §§ 11021(b), 11022(b); 40 C.F.R. § 370.20(b).

55/ 42 U.S.C. § 11023; 40 C.F.R. pt. 372.

56/ 42 U.S.C. § 11045.

enforcement actions brought by citizens as well as state and local governmental units.57/

8.10.2 Iowa Community Right-To-Know

Iowa's Hazardous Chemicals Risks Right-To-Know Act establishes a public right to be informed about the presence of hazardous chemicals in the community and the potential health and environmental hazards that the chemicals pose.58/ As with employee right-to-know, the Labor Commissioner administers the community right-to-know law. In particular, the Labor Commissioner will provide annually to the main county seat library a list of facilities and chemicals reported on Tier Two forms under EPCRA.59/ An employer must submit to the local fire department a list of hazardous chemicals that are consistently at or transported from the employer's facility.60/

Under the Labor Commissioner's rules, an employer has a duty to inform the public, upon request, of the presence of hazardous chemicals in the community and the potential health and environmental hazards that the chemicals pose.61/ An interested person may request information from an employer. The employer shall provide the information or reason for refusal within ten days. If the request is denied, the requesting party may file an application for review with the Labor Commissioner. The Labor Commissioner has authority to order the employer to comply, and administrative review of such an order is available under the Iowa Administrative Procedure Act.62/

57/ 42 U.S.C. § 11046.

58/ Iowa Code § 89B.12.

59/ Iowa Admin. Code § 347-130.12.

60/ Iowa Code § 89B.15; Iowa Admin. Code § 347-140.5.

61/ Iowa Admin. Code § 347-130.1.

62/ Iowa Admin. Code § 347-130.10.

8.11 Enforcement

A person who fails to notify the DNR and the local police department or the office of the sheriff of the affected county of the occurrence of a hazardous condition is subject to a civil penalty of not more than one thousand dollars.63/

An owner, operator, or person in charge of a vessel, onshore facility, or offshore facility that improperly discharges oil or a hazardous substance may be fined under the Clean Water Act as amended by OPA by a class I or class II civil administrative penalty.64/ A class I civil administrative penalty may not exceed $10,000 per violation or $25,000 total, and a class II civil administrative penalty may not exceed $10,000 per day of violation or $125,000 total. As an alternative to the administrative penalty, such a violation may be subject to a civil penalty of $25,000 per day of violation or up to $1,000 per barrel of oil or unit of reportable quantity of hazardous substances discharged.65/ Violations that result from gross negligence are subject to a civil penalty not less than $100,000 and not more than $3,000 per barrel of oil or unit of reportable quantity of hazardous substance discharged.66/

RCRA hazardous waste and petroleum underground storage tank notification regulations are backed by civil enforcement authority of $25,000 per day per violation plus criminal enforcement authority set forth in 42 U.S.C. § 6928(a)(3). PCB spill regulations are backed by civil enforcement authority of $20,000 per day per violation plus criminal enforcement authority at 15 U.S.C. § 2615. Knowing violations of the Hazardous Materials Transportation Act may trigger civil penalties up to $25,000 per day of violation.67/ Willful violations of the Hazardous Materials

---

63/ Iowa Code § 455B.386.

64/ 33 U.S.C. § 1321(b)(6).

65/ 33 U.S.C. §§ 1321(6)(E), (7)(A).

66/ 33 U.S.C. § 1321(7)(D).

67/ 49 U.S.C. § 5123.

Transportation Act may result in criminal penalties including imprisonment for not more than five years.[68]

OSHA is backed by civil enforcement authority of up to $70,000 per violation plus criminal enforcement authority set forth in 29 U.S.C. § 666. EPCRA is backed by civil enforcement authority of $25,000 per day per violation plus criminal enforcement authority set forth in 42 U.S.C. § 11045.

68/ 49 U.S.C. § 5124.

CHAPTER NINE

SOLID WASTE MANAGEMENT

9.0 Introduction

This chapter discusses Iowa's laws and regulations governing solid waste disposal. Although historically the most common method for disposing of solid waste has been the open dump or landfill, Iowa has put forward a comprehensive scheme of solid waste management to avoid some of the problems associated with landfill disposal of waste such as groundwater and other environmental contamination. Iowa also has adopted progressive recycling requirements.

9.1 Overview Of Iowa Solid Waste Management Laws And Policy

9.1.1 Solid Waste Defined

Iowa law defines "solid waste" as:

> [G]arbage, refuse, rubbish, and other similar discarded solid or semisolid materials, including but not limited to such materials resulting from industrial, commercial, agricultural, and domestic activities. Solid waste may include vehicles as defined by section 321.1, subsection 90. However, this division does not prohibit the use of dirt, stone, brick, or similar inorganic material for fill, landscaping, excavation or grading at places other than a sanitary disposal project. Solid waste does not include hazardous waste as defined in section 455B.411 or source, special nuclear, or by-product material as defined in the Atomic Energy Act of 1954, as amended to January 1, 1979, or petroleum contaminated soil which has been remediated to acceptable state or federal standards.1/

9.1.2 RCRA Regulation Of Solid Waste

The federal Resource Conservation and Recovery Act (RCRA) governs solid waste management as well as hazardous waste management, which is discussed in chapter 6 of this handbook. RCRA prohibits the disposal of solid waste in "open

---

1/ Iowa Code § 455B.301(20).

dumps."[2] EPA regulations adopted under RCRA establish minimum criteria for municipal solid waste (MSW) landfills to determine whether the facility is classified as an "open dump" or not.[3] In general, however, RCRA subtitle D provides that states are primarily responsible for management of non-hazardous solid waste.

9.1.3 Iowa Regulation Of Solid Waste

The State of Iowa recognizes that landfills are a necessary part of solid waste disposal but encourages alternative methods of solid waste management to protect the environment and the public and to promote the practical and beneficial use of material and the potential energy value of solid waste.[4] To further this purpose, Iowa has established a state "solid waste management policy" which outlines the following waste management hierarchy in descending order of preference:

1. Volume reduction at the source;
2. Recycling and reuse; and
3. Other approved techniques of solid waste management including, but not limited to, combustion with energy recovery, combustion for waste disposal, and disposal in sanitary landfills.[5]

To implement this solid waste policy, Iowa law directs the DNR to establish and maintain a cooperative state and local program of project planning and technical and financial assistance to encourage comprehensive solid waste management.[6]

9.2 Iowa Department of Natural Resources and Regulation of Solid Waste

The Iowa DNR is the state agency with delegated authority to administer Iowa solid waste laws. The DNR director is responsible for sanitary disposal project

---

2/ 42 U.S.C. § 6945; 40 C.F.R. § 256.23.

3/ 40 C.F.R. § 258.1(g); see also 40 C.F.R. pt. 241 (guidelines for land disposal of solid waste).

4/ Iowa Code § 455B.301A(1).

5/ Iowa Code § 455B.301A(1).

6/ Iowa Code §§ 455A.4, 455B.303.

permits, reviewing city and county comprehensive plans, and the overall supervision of solid waste management in the state.7/ The Iowa Environmental Protection Commission (EPC) establishes the policy for the DNR and adopts rules necessary for the effective administration of the Iowa solid waste laws.8/ Cities and counties also play a role in developing and implementing Iowa's solid waste management policy at the local level.

Solid waste may only be disposed of in "sanitary disposal projects," which are subject to numerous permitting and operating laws and regulations. The EPC has adopted specific solid waste-related rules for sanitary disposal projects, sanitary landfills, sanitary disposal projects with processing facilities, yard waste disposal and solid waste composting facilities, and recycling operations.

9.3 Local Governments and Regulation of Solid Waste; Comprehensive Plans

Under Iowa's solid waste policy, cities and counties must establish a comprehensive solid waste reduction program consistent with the state waste management policy and a comprehensive project for final disposal of waste.9/ Cities and counties are eligible for state grants to implement the state's solid waste management policy.10/ Cities and counties may establish such programs separately or may work together to establish joint projects for their citizens.11/ All cities and counties or Chapter 28E agencies (private individuals or business organizations under contract with a local government)12/ must file with the DNR director a

---

7/ Iowa Code § 455A.4.

8/ Iowa Code § 455A.6.

9/ Iowa Code § 455B.302; Iowa Admin. Code § 567-101.4 to 101.5.

10/ Iowa Code § 455B.311.

11/ Iowa Code § 455B.302.

12/ See Iowa Code § 28E.2 ("The term 'private agency' shall mean an individual and any form of business organization authorized under the laws of this or any other state"); Iowa Code § 28E.4 ("Any public agency of this state may enter into an agreement with one or more public or private

comprehensive plan detailing the method of solid waste reduction for their residents. Cities and counties working together on such programs may file a single plan.13/

Comprehensive plans must include the following:

1. A description of the planning area and agencies involved;
2. A description of past local and regional planning activities;
3. Waste stream information;
4. A 20-year projection of waste composition and waste generation rates;
5. A description of the existing waste management system;
6. An analysis of alternative waste management systems in light of the state's waste management hierarchy;
7. A description of the proposed waste management system in light of the alternatives;
8. A specific plan and schedule for implementing the comprehensive plan no later than July 1, 1997; and
9. A description of the proposed methods of financing.14/

The role of local governments in siting sanitary disposal projects is discussed in section 9.5 of this handbook.

9.4 Sanitary Disposal Project Permits

A sanitary disposal project consists of the facilities and equipment to "facilitate the final disposal of solid waste without creating a significant hazard to the public health or safety."15/ Subject to certain exceptions, it is unlawful to deposit

---

agencies for joint or co-operative action . . . . Appropriate action by ordinance, resolution or otherwise pursuant to law of the governing bodies involved shall be necessary before any such agreement may enter into force.").

13/ Iowa Code § 455B.306.

14/ Iowa Admin. Code § 567-101.5(4).

15/ Iowa Code § 455B.301(18).

or dispose of solid waste anywhere other than at a sanitary disposal project.16/ In order to construct or operate a sanitary disposal project, a person or agency must apply for and receive a permit from the DNR director.17/ Among other items, the application must include:

1. The name, address and phone number of the owner, applicant, and others;
2. A legal description of the site and an aerial photograph of the site;
3. The type, source, and expected volume of waste on a daily, weekly, or yearly basis;
4. An organizational chart;
5. A detailed description of the disposal process to be used;
6. A table listing the equipment to be used, its design capacities, and expected loads;
7. Contingency plan to be used in case of equipment breakdown, fire, or other emergencies;
8. Proof of the applicant's site ownership; and
9. A closure and postclosure plan which includes the applicant's groundwater monitoring plans and leachate control system.18/

The applicant for an initial permit or a renewal permit must submit a comprehensive plan developed in cooperation with the city or county responsible for establishing the sanitary disposal project (if a private entity is submitting

---

16/ Iowa Code § 455B.307(1). However, this prohibition does not by itself grant authority to the DNR to compel a landowner to clean up already existing solid waste on his or her property. See First Iowa State Bank v. Iowa DNR, 502 N.W.2d 164 (Iowa 1993); see also State v. Shelley, 512 N.W.2d 579 (Iowa App. 1993) (upholding DNR order requiring owners to cease disposal of solid waste on their property, remove solid waste, and pay $1,000 penalty based on owners' failure to challenge order within 30 days); State v. Grell, 368 N.W.2d 139 (Iowa 1985) (holding that salvage is solid waste under Iowa Code § 455B.307).

17/ Iowa Admin. Code § 567-102.1.

18/ Iowa Admin. Code § 567-102.12.

application).[19] The comprehensive plan must reflect and incorporate the state's solid waste management hierarchy and solid waste management policy or else the responsible city or county must hold a public hearing to address the basis for not including such elements in the plan.[20]

As of July 1, 1994, the Iowa legislature prohibited the DNR director from renewing or reissuing permits unless the permit applicant has taken all steps in conjunction with local governments to implement the comprehensive plan.[21] Prior to approving plans for a new sanitary disposal project, the local governing body where the landfill will be located must hold a public hearing to address whether to include curbside recycling in the applicant's comprehensive plan.[22]

Permits for sanitary disposal projects are effective for three years with a right to apply for renewal at that time.[23] In addition to the three-year permits for sanitary disposal projects, the DNR may issue temporary permits for one-year renewable terms and developmental permits for one to three-year renewable terms.[24] These latter permits are issued to sites which cannot comply with the rules of a sanitary disposal project but will serve the public interest and will not cause adverse health-related and environmental effects. Finally, the DNR issues closure permits for facilities that no longer accept solid waste. These permits are issued for 30-year renewable terms which remain in effect until the DNR determines that postclosure maintenance, monitoring, and operation of leachate control systems are no longer necessary to protect surrounding groundwater.[25]

19/ Iowa Code § 455B.306(1).

20/ Iowa Code § 455B.306(4).

21/ Iowa Code § 455B.306(6).

22/ Iowa Code § 455B.306(10).

23/ Iowa Admin. Code § 567-102.2(1).

24/ Iowa Admin. Code § 567-102.2(2), (3).

25/ Iowa Admin. Code § 567-102.2(4).

Additional information is required for applicants seeking permits for certain types of sanitary disposal projects such as sanitary landfills, sanitary disposal projects with processing facilities, yard waste and solid waste composting facilities, and recycling operations. For example, applicants for a sanitary landfill permit must submit a scale drawing map; detailed engineering drawings of the site; a description of the proposed landfill liner including the method of installation; diversion and drainage structures to prevent ponding and erosion; a leachate collection, storage and treatment, and disposal system to prevent surface water and groundwater from leachate contamination;[26] plans for a hydrogeologic monitoring system;[27] and a groundwater quality assessment plan.[28]

As of July 1, 1997, the director may not renew or reissue a sanitary landfill permit unless the permit applicant, in conjunction with all local governments, has documented alternative methods of solid waste disposal other than landfill disposal.[29] Presently, the director may not issue new sanitary landfill permits unless the landfill is equipped with a leachate collection system,[30] and may not renew or reissue permits for existing landfills unless the landfill is equipped with a leachate collection system.[31]

Rules relating to requirements for other types of disposal projects are as follows:

26/ Iowa Admin. Code § 567-103.2(11).

27/ Iowa Admin. Code § 567-103.2(5).

28/ Iowa Admin. Code § 567-103.2(9).

29/ Iowa Code § 455B.305(5).

30/ Iowa Admin. Code § 567-103.2(11).

31/ Iowa Code § 103.2(12).

- Sanitary disposal projects with processing facilities -- Iowa Admin. Code § 567-104
- Sanitary landfills accepting only construction and demolition waste -- Iowa Admin. Code § 567-103.4
- Sanitary landfills accepting only specific types of solid waste -- Iowa Admin. Code § 567-103.5
- Sanitary landfills accepting only municipal sewage sludge -- Iowa Admin. Code § 567-103.6
- Yard waste disposal and solid waste composting facilities -- Iowa Admin. Code § 567-105

A facility permit may be suspended or revoked by the DNR director if the sanitary disposal project fails to meet any of the requirements of the solid waste laws or rules.[32] Any adverse action by the director may be appealed to the DNR.[33]

9.5 Siting of Sanitary Disposal Projects

Local governments play an important role in connection with the location of sanitary disposal projects. Prior to the siting of a proposed new sanitary landfill, incinerator, or infectious medical waste incinerator, the applicant must submit a request for local siting approval to the city council or board of supervisors which governs the city or county containing the proposed site. The applicant for siting approval must demonstrate compliance with all permit requirements for the proposed facility as well as show the following:

1. The project is necessary to accommodate the needs of the area;
2. The project is designed, located, and proposed to be operated in a way that protects public health, safety, and welfare;
3. The project attempts to minimize incompatibility with the character of The surrounding area and the effect on surrounding property values;
4. The project attempts to minimize danger to the surrounding area from fire, spills, and other operational accidents;

32/ Iowa Code § 455B.305(1).

33/ Iowa Code § 455B.308.

5. The project attempts to minimize impact of existing traffic patterns and traffic flows;
6. The applicant's previous operating experience has been made available to the governing local body; and
7. The governing local body has consulted the DNR prior to approval.[34]

No later than fourteen days prior to the request for siting approval, the applicant must give written notice of the request for siting to all property owners within 1,000 feet of the lot line if the facility is within city limits or 2,000 feet of the lot line if the facility is outside city limits. The applicant must also publish written notice of the request in the official county newspaper.[35]

Anyone may file written comments with the governing local body concerning the appropriateness of the proposed facility, and the governing body shall consider any comments received not later than thirty days after the date of the last public hearing. No sooner than 90 days before and no later than 120 days after receipt of the request for siting approval, the governing local body shall hold at least one public hearing on the proposed siting.[36] Siting decisions must be in writing and issued within 180 days from the request.[37] If approval is granted, the project must begin within one year of the approval date.[38]

9.6 Sanitary Disposal Project Operating and Monitoring Requirements

The operation of a sanitary disposal project is subject to extensive regulation. Sanitary disposal projects may not accept radioactive waste and, except with a permit and instructions from the DNR, sanitary disposal projects may not accept hazardous,

---

34/ Iowa Code § 455B.305A(2).

35/ Iowa Code § 455B.305A(3).

36/ Iowa Code § 455B.305A(5).

37/ Iowa Code § 455B.305A(6).

38/ Iowa Code § 455B.305A(7).

toxic, or infectious waste; sewage sludge; or waste tires.39/ In addition, as of January 1, 1991, land disposal of yard waste is prohibited40/ and as of January 1, 1992, disposal of baled solid waste is prohibited.41/

Iowa rules regulate the operation of sanitary disposal projects and include restrictions on open burning, litter containment, scavenging, insect and rodent control, maintenance of internal roads, and the control of free liquid waste.42/ Additional regulations exist for the operation of sanitary landfills. These rules cover restrictions on loading and unloading of waste, site access, inspections, grading, the operation of the landfill's hydrologic monitoring system (for the purpose of measuring the impact of the landfill on the surrounding groundwater), proper laboratory procedures for groundwater sampling, proper recordkeeping and recording, and control of explosive gases.43/

Specific hydrologic monitoring standards exist for all solid waste disposal facilities. These rules address proper soil investigation, groundwater level measurements, reporting requirements, evaluation of hydrogeologic conditions, sampling protocol, and construction standards for monitoring wells and soil borings.44/

Additional operating requirements apply to the following facilities:

39/ Iowa Admin. Code § 567-102.15.

40/ Iowa Code § 455D.9.

41/ Iowa Code § 455D.9A.

42/ Iowa Admin. Code § 567-102.14.

43/ Iowa Admin. Code § 567-103.2.

44/ Iowa Admin. Code § 567-110.

- Sanitary landfills accepting only construction and demolition waste -- Iowa Admin. Code § 567-103.4(2)
- Sanitary landfills accepting only specific types of waste -- Iowa Admin. Code § 567-103.5(2)
- Sanitary landfills accepting only municipal sewage sludge -- Iowa Admin. Code § 567-103.6(2)
- Sanitary disposal projects with processing facilities -- Iowa Admin. Code § 567.104.10
- Yard waste composting facilities -- Iowa Admin. Code § 567.105.3; Iowa Admin. Code § 567.105.4
- Solid waste composting facilities -- Iowa Admin. Code § 567-105.3; Iowa Admin. Code § 567-105.6

All sanitary landfill operators and solid waste incinerator operators must be trained, tested, and certified by a DNR-approved certification program.45/ The basic operator training course must include at a minimum 25 contact hours and cover several areas which include information on types of waste, design, monitoring, and landfill and incinerator operations.46/

9.7 Closure Requirements For Sanitary Disposal Projects

Iowa strictly regulates the closure and postclosure procedures for sanitary disposal projects and sanitary landfills in particular. Such regulation is necessary to ensure that landfill owners continue to monitor effectively for potential groundwater contamination, which may occur many years after the landfill ceases operation. Owners or operators of sanitary disposal projects must notify the DNR in writing at least 180 days prior to closure of the facility or suspension of operations and must implement a closure/postclosure plan within 90 days of closure.47/

With regard to the closure of sanitary landfills, a permit applicant shall include in its comprehensive plan a closure and postclosure plan detailing the schedule for and methods by which the operator will meet all closure conditions.

45/ Iowa Admin. Code § 567-102.13.

46/ Iowa Admin. Code § 567-102.13(4).

47/ Iowa Admin. Code § 567-102.14(9).

The plan must include the proposed frequency and types of action to be implemented before and after closure, the proposed postclosure actions to be taken to return the area to a condition suitable for other uses, and a cost estimate. The postclosure plan must reflect a thirty-year time period for postclosure responsibilities.[48]

For thirty years following closure of the landfill, the owner or operator of the site must comply with all postclosure requirements which include upkeep of the final cover; maintenance of proper drainage; continued maintenance and operation of the groundwater monitoring system, leachate collection system, and gas monitoring and collection system; and submitting semiannual reports to the DNR containing information about general conditions at the site and all monitoring results.[49]

9.8 Financial Responsibility For Sanitary Disposal Projects

As part of the closure planning effort, owners and operators of sanitary disposal projects must demonstrate and maintain financial assurance for site closure, postclosure care, and contingency remedial action. Persons or entities applying for a sanitary disposal project permit must submit a financial assurance instrument prior to initial approval of the permit or to the renewal of an existing permit.[50] Financial assurance instruments may include cash or surety bonds, a letter of credit, a secured trust fund, or a corporate guarantee.[51] The financial assurance instrument shall not be canceled, revoked, disbursed, released, or allowed to terminate without DNR approval. This obligation continues after the project ceases to operate until the operator is released from closure, postclosure, and

---

48/ Iowa Code § 455B.306(6)(a).

49/ Iowa Admin. Code § 567-103.2(14).

50/ Iowa Code § 455B.306(8).

51/ Iowa Code § 455B.306(8)(d); <u>see also</u> Iowa Code § 455B.301(8) (definition).

monitoring responsibilities.52/ In addition, the operator must submit annual financial statements to the DNR which include the current amount in the financial assurance instrument.53/

9.9 Managing Special Solid Wastes

9.9.1 White Goods Disposal

Certain large home appliances or "white goods" such as refrigerators, freezers, washers, dryers, and microwave ovens may contain polychlorinated biphenyls (PCBs) which are potentially hazardous to the environment. As a result, Iowa law provides that all white goods must be inspected for the presence of PCB-containing capacitators and that all PCB-containing capacitors must be removed from white goods prior to processing and disposed of by placing them in DOT-approved containers which are sealed, labeled with an EPA-approved PCB labels, and disposed of at EPA-approved waste disposal facilities for PCBs.54/

9.9.2 Land-Farming

Subject to limitations, certain solid wastes may be applied to land or "land-farmed." Under most circumstances, a permit from the DNR is required for land application of waste, but land application is authorized without a permit at certain levels of thickness and levels of toxic constituents for municipal sewage sludge and petroleum-contaminated soils.55/ All other land application of waste requires a permit.56/

52/ Iowa Code § 455B.306(8)(a).

53/ Iowa Code § 455B.306(6)(c) and (8)(e).

54/ Iowa Admin. Code. § 567-118.3(2) and (3); see also chapter 13 of this handbook regarding PCB regulation.

55/ Iowa Admin. Code §§ 567-121.3(1), 121.3(2).

56/ Iowa Admin. Code § 567-121.4.

9.9.3 Bottles And Other Containers

Iowa places volume reduction and recycling at the top of the state's waste management hierarchy and has adopted policies and regulations which promote government, private industry, and citizen involvement in recycling efforts. In 1981, Iowa passed a "bottle bill" under which all beverage containers (sealed glass, plastic, or metal bottles, cans, jars, or cartons containing nearly all alcoholic and nonalcoholic drinks for human consumption) are subject to a minimum deposit of five cents.57/ All dealers must accept empty beverage containers from and pay the refund value to consumers for all beverage containers that the dealer sells unless the dealer is covered by an approved redemption center.58/ Since July 1, 1990, it has been illegal for a dealer, distributor, manufacturer of beverage containers, or a person operating a redemption center to dispose of most beverage containers in sanitary landfills.59/

9.9.4 Ash, Foundry Sand, And Other Wastes

In addition, Iowa rules specify the conditions under which certain solid waste may be reused without a solid waste permit.60/ In particular, coal combustion residue may be stored or used as a raw material in cement or concrete or as a fill base for roads or parking lots,61/ and certain types of used foundry sand may be used for landfill cover, construction fill, or other beneficial uses.62/

57/ Iowa Code §§ 455C.1(2), 455C.2; Iowa Admin. Code § 567-107.1.

58/ Iowa Code § 455C.3; Iowa Admin. Code § 567-107.4(1).

59/ Iowa Code § 455C.16.

60/ Iowa Admin. Code § 567-108.1.

61/ Iowa Admin. Code § 567-108.4(3).

62/ Iowa Admin. Code § 567-108.4(4); see also West Bend Mut. Ins. Co. v. Iowa Iron Works, Inc., 503 N.W.2d 596 (Iowa 1993) (involving insurance coverage for alleged violations of Iowa Code § 455B.307 and Iowa Admin. Code § 567-101.1, 101.3(1) based on foundry sand disposal).

9.10 Waste Volume Reduction and Recycling Act

In 1989, Iowa passed the Waste Volume Reduction and Recycling Act.[63] The Act directs the DNR to establish a statewide waste reduction and recycling network which includes efforts to promote the following:

1. Increase the amounts of recyclable materials used by the public;
2. Recover recyclable materials from the waste stream;
3. Promote local efforts to implement recycling collection centers located at disposal sites or other local sites;
4. Promote local efforts of curbside collection of separated recyclable waste materials;
5. Provide public education programs which promote awareness of waste reduction and recycling;
6. Promote the creation of markets for recyclable materials;
7. Promote research, manufacturing processes, and product development which provide for waste reduction; and
8. Promote certain business, industry, and academic research programs.[64]

The Act also prohibits the disposal of certain types of waste at sanitary landfills. In some cases (such as batteries), the manufacturers or retailers must take back the batteries from consumers,[65] and in other cases (yard waste), the Act encourages the local establishment of composting and collection facilities.[66] The

---

63/ Iowa Code ch. 455D.

64/ Iowa Code § 455D.5(1); see also chapter 17 of this handbook for a discussion of pollution prevention.

65/ Iowa Code § 455D.10.

66/ Iowa Code § 455D.9.

Act prohibits disposal in landfills of waste tires prior to proper processing, waste oil, and baled solid waste.[67]

State procurement laws require state offices to purchase recycled goods where possible and practical.[68] All state agencies must establish wastepaper recycling programs.[69]

9.11 Packaging Restrictions

Iowa has adopted laws aiming to decrease the amount of packaging material that is potentially harmful to the environment. The DNR has established a recycling program in cooperation with businesses involved in the manufacture and use of polystyrene (PS) packaging products or food service items to increase the recycling of such products or items by 25 percent by July 1, 1993, and by 50 percent by July 1, 1994.[70] If these deadlines are not met, the manufacture, sale, or use of any polystyrene packaging products or food service items will be banned in Iowa after January 1, 1996.[71] At the time of this writing, the DNR continues to evaluate this issue.

Iowa statutes based on the Coalition of Northeastern Governors (CONEG) model toxics proposal regulate and prohibit the use of some packages and packaging components containing the heavy metals lead, cadmium, hexavalent chromium, and mercury which may contaminate the environment upon incineration or

---

67/ Iowa Code §§ 455D.9A, 455D.11; Iowa Admin. Code § 567-119.3; see also Ervin v. Iowa Dist. Ct. for Webster County, 495 N.W.2d 742 (Iowa 1993) (upholding contempt order imposing 6 months imprisonment stayed after 20 day for violation of waste tire restrictions and court order relating to same).

68/ Iowa Code §§ 18.6, 18.18, 18.21, 262.9, 307.21, 904.312; Iowa Admin. Code § 567-119.8.

69/ Iowa Code § 18.20.

70/ Iowa Code § 455D.5(3).

71/ Iowa Code § 455D.16.

landfill disposal.[72] With few exceptions, the concentration levels of lead, cadmium, mercury, or hexavalent chromium present in any package or packaging component may not exceed 600 parts per million (ppm) effective July 1, 1992, 250 ppm effective July 1, 1993, and 100 ppm effective July 1, 1994. Violations of this legislation are considered a misdemeanor.

After July 1, 1992, rigid plastic containers and plastic bottles must be coded as to resin content.[73] For this coding, Iowa has adopted the Society of the Plastics Industry (SPI) voluntary resin identification symbol system that is required in many states; i.e. number 1 is polyethylene terephthalate (PETE), etc. A container manufacturer or distributor who violates these provisions is subject to a civil penalty not to exceed $500 for each violation.

9.12 Waste Designation or Flow Control

In recent years, a number of states and local governments around the country have experimented with waste designation or "flow control" legislation which restricts the flow of waste into or out of a particular state, county, or city. Although governments historically attempted to keep waste out of their jurisdiction,[74] more recently, state and local governments have expended significant resources to construct and operate waste processing facilities and have enacted corresponding legislation to ensure a sufficient stream or flow of solid waste to finance these facilities through "tipping fees." Because such legislation often curtails the business of local waste haulers and landfills, flow control legislation has led to a significant amount of litigation.

In 1983, the United States Court of Appeals for the Eighth Circuit reviewed an antitrust claim brought by a waste collection and disposal service against the Des

---

72/ Iowa Code § 455D.19.

73/ Iowa Code § 455D.12.

74/ See Philadelphia v. New Jersey, 437 U.S. 617 (1978).

Moines Metropolitan Area Solid Waste Agency, its constituent counties, and numerous municipalities. The plaintiff claimed that antitrust laws prohibited the Agency from requiring that all solid waste generated within the geographic area covered by the Agency be disposed of at the Agency's facilities.75/ The Eighth Circuit rejected this antitrust challenge to flow control.

In May 1994, the U.S. Supreme Court held in Carbone v. Town of Clarkstown76/ that state and local laws requiring waste generated within a particular jurisdiction to be processed or disposed of at a designated local facility violated the U.S. Commerce Clause. Consistent with Carbone, the United States Court of Appeals for the Eighth Circuit in Waste Systems Corp. v. County of Martin held that county ordinances requiring all waste generated in Martin and Faribault counties in Minnesota to be delivered to county-owned facilities for processing discriminated against interstate commerce.77/ Congress is presently considering a bill that would allow states and local governments to override commerce clause restrictions and enact designation laws under some circumstances.

9.13 Superfund Litigation Involving Municipal Solid Waste

As noted above, municipal solid waste (MSW) is not regulated under RCRA or Iowa law as a hazardous waste. Based in part on the distinction between

---

75/ Central Iowa Refuse Systems, Inc. v. Des Moines Metro. Solid Waste Agency, 715 F.2d 419 (8th Cir. 1983) (holding that Iowa law encouraged local governments to provide for environmentally sound disposal of solid waste and allowed the use of revenue bonds to build appropriate waste facilities resulting in application of state action exception to antitrust laws), cert. denied, 471 U.S. 1003 (1985).

76/ 114 S. Ct. 1677 (1994).

77/ 985 F.2d 1381 (8th Cir. 1993). In a similar vein, the U.S. Supreme Court has struck down state and local laws which charge higher disposal and processing fees for out-of-state or out-of-county waste as violating the Commerce Clause. See Oregon Waste Sys., Inc. v. Oregon DEQ, 114 S. Ct. 1345 (1994) (striking down Oregon statute that imposed additional fee on solid waste generated outside state and disposed of within state); Chemical Waste Management v. Hunt, 112 S. Ct. 2009 (1992) (striking down Alabama statute that imposed additional fee on hazardous waste generated outside state and disposed of within state).

"hazardous waste" and "hazardous substances," however, municipalities and other parties involved in operating MSW dumps or landfills, MSW collection, and MSW hauling have been sued under the Superfund law seeking recovery of the costs of cleaning up hazardous substances at old dump sites.78/

The EPA adopted a policy which provides that "when the source of the municipal waste is believed to come from households, regardless of whether household hazardous waste may be present, the general policy is to exclude such municipal wastes from the Superfund process."79/ A New Jersey federal district court rejected a challenge under the equal protection clause of the United States Constitution to EPA's policy of not pursuing municipal generators or transporters of MSW under CERCLA.80/ Private parties, however, are not bound by the EPA policy.

At about the same time, defendants to an EPA lawsuit filed a third-party action in federal court in Connecticut against numerous third-party defendants including several municipalities81/ that had collected and dumped garbage mostly in the 1970s and 1980s. The United States Court of Appeals for the Second Circuit ruled that the municipalities may be liable for response costs under the Superfund law.82/

78/ See chapter 5 of this handbook discussing Superfund liability.

79/ EPA, Superfund Program; Interim Municipal Settlement Policy, 54 Fed. Reg. 51,071, at 51,072 (Dec. 12, 1989).

80/ United States v. Helen Kramer, 757 F. Supp. 397 (D. N.J. 1991).

81/ B. F. Goodrich Co. v. Murtha, 754 F. Supp. 960 (D. Conn. 1991), aff'd, 958 F.2d 1192 (2d Cir. 1992).

82/ After a more recent rehearing, the Connecticut district court found on December 20, 1993 that the parties could not prove that the MSW contained a sufficient amount of hazardous substances to result in cleanup costs and the court dismissed the claims against the municipalities.

Similar cases have been filed around the country.[83] EPA has proposed legislation that would limit the Superfund liability of municipal generators and transporters to 10% of the cleanup costs at former landfills in most circumstances. In the meantime, Superfund liability for MSW disposal continues to be litigated with uncertain outcomes.

83/ See Transportation Leasing Co. v. California, No. CV-89-7368-WNB (C.D. Cal. Dec. 4, 1990) (municipality may be liable if plaintiffs "prove that the waste disposed of at the [facility] contained 'hazardous substances' under CERCLA"), subsequent proceedings at, 1993 Westlaw 733,014 (C.D.Cal. 1993); Anderson v. City of Minnetonka, 1993 Westlaw 95,361 (D. Minn. Jan. 27, 1993) (municipal solid waste disposal may trigger CERCLA liability).

CHAPTER TEN

STORAGE TANK REGULATION

10.0 Introduction To UST Regulation

Regulatory programs administered by the Iowa Department of Natural Resources govern underground storage tank (UST) installation and maintenance to prevent, control, and clean up leaks. Iowa also has a programs designed to reimburse qualifying owners and operators for part of their costs of cleaning up UST leaks and to enable UST owners and operators to obtain insurance to cover the costs of cleaning up future leaks. DNR estimates that there are 12,395 UST sites in Iowa. A less extensive regulatory program for petroleum aboveground storage tanks (ASTs) is administered by the office of the Iowa State Fire Marshal.

With regard to federal law, the 1984 amendments to the Resource Conservation and Recovery Act (RCRA) authorized the EPA to adopt regulations governing USTs.[1] The Environmental Protection Commission of the DNR has adopted the federal regulations essentially verbatim with additional requirements applicable to farm and residential tanks smaller than 1,110 gallons. In 1994, Iowa applied for EPA approval of its UST program to operate in lieu of the federal program, specifying that Iowa DNR does not seek authority over Indian lands. Because EPA has published notice of its intent to approve Iowa's application,[2] as a practical matter DNR administers most tank regulations in Iowa.

10.1 Which USTs Are Covered, And Who Is Responsible For Them?

Iowa UST regulations apply to tanks and associated piping if 10 percent of the volume of the system is below the surface of the ground and the system contains a "regulated substance."[3] "Regulated substance" is broadly defined to include any

---

1/ 42 U.S.C. §§ 6991-6991i; 40 C.F.R. pt. 280.

2/ 59 Fed. Reg. 40,507 (Aug. 9, 1994).

3/ Iowa Code § 455B.471(11).

substance that could create a "substantial danger" to the public if released.[4] However, any substance which is regulated under the RCRA Subtitle C provisions relating to the storage, transport, or disposal of hazardous wastes discussed in chapter 6 of this handbook (RCRA hazardous waste) is exempt from UST regulation. The vast majority of USTs are "petroleum USTs" (including gasoline, heating oil, and jet fuel USTs). "Petroleum USTs" are treated differently in some instances than "hazardous substance" USTs, which include USTs containing any regulated substance identified as a hazardous substance under the federal Superfund Act discussed in chapter 5 of this handbook. Examples of hazardous substance USTs include tanks containing industrial solvents or pesticides.

The following types of USTs are generally exempt from UST regulation:

1. Home heating oil tanks;
2. Residential septic tanks;
3. Flow-through process tanks;
4. Tanks with a capacity of less than 110 gallons;
5. Pipelines regulated under other federal laws;[5]
6. Components of regulated wastewater treatment systems;
7. Surface impoundments, pits, ponds, or lagoons;
8. Liquid traps or associated gathering lines directly related to oil or gas production and gathering operations; and
9. Tanks located below ground level, but above a floor (i.e., in a basement).[6]

4/ Iowa Code § 455B.471(8).

5/ See, e.g., 49 U.S.C. § 2000; Iowa Code ch. 479; see also Kinley Corp. v. Iowa Util. Bd., 999 F.2d 354, 359 (8th Cir. 1993) (federal hazardous liquid pipeline act preempts Iowa Code ch. 479 with regard to jet fuel pipeline); ANR Pipeline Co. v. Iowa State Commerce Comm'n, 828 F.2d 465 (federal act preempts Iowa Code ch. 479 regarding natural gas pipeline).

6/ Iowa Code § 455B.471(11); Iowa Admin. Code §§ 561-135.1(3), 561-135.2.

Farm or residential USTs with a capacity less than 1,100 gallons used for storing motor fuel for noncommercial purposes also are exempt from most UST regulations if they were installed before July 1, 1987 (although these tanks are nonetheless subject to registration requirements). Farm and residential tanks meeting this description that were installed after July 1, 1987 are subject to all petroleum UST regulations, although owners do not have to comply with "financial responsibility" rules.

Generally, USTs regulations place responsibility for installing and maintaining a UST system, and for cleaning up any spills, on the "owner or operator" of a UST. If a person falls within the definition of owner of operator, he or she cannot eliminate his liability by contract. For example, if a gas station owner leases the station to an independent operator, and the operator agrees to take full responsibility for the USTs, the owner remains liable under the UST laws if the operator fails to live up to their private contract.7/ Lenders who hold a security interest in a property with a UST or foreclose on such a property may be exempt from the definition of "owner" if they do not exercise managerial control over the property.8/

10.2 UST Registration And Proof Of Financial Responsibility

With a few exceptions discussed below, USTs must be registered with the State annually. By January 15 of each year or within thirty days of completing a new installation, the UST owner or operator must file with the DNR a completed Tank Registration Form (DNR Form 148) and pay a registration fee (currently $10 per

---

7/ Cf. In re Novak, DIA No. 90DNR-14, 1990 Westlaw 207,513 (Sept. 17, 1990) (service station buyer's release of seller's liability "cannot be enforced" against the DNR).

8/ Iowa Code § 455B.471(6)(b); see also chapter 16 of this handbook for further discussion of lender liability. Pursuant to 42 U.S.C. § 6991b(h)(9), the EPA has proposed similar rules regarding lender liability relating to USTs. 59 Fed. Reg. 30,448 (June 13, 1994) (to be codified at 40 C.F.R. §§ 280.200-280.250, 281.39).

tank) and a tank management fee (currently $65 per tank). Most of the tank management fee is used for the cleanup cost reimbursement programs discussed below as well as to fund cleanup actions taken by DNR when no responsible party can be identified. Upon payment of the these fees, the DNR will issue an annual registration tag that must be attached to the fill pipe of each registered UST. Anyone who fills a UST that does not have a current registration tag on it may be personally liable.9/

It is not necessary to register tanks taken out of service and closed before January 1, 1974, removed from the ground before July 1, 1985, or closed in compliance with UST closure rules (discussed below) after these dates. Farm or residential USTs with a capacity less than 1,100 gallons used for storing motor fuel for noncommercial purposes existing before July 1, 1987, were required to be registered by July 1, 1989. These tanks are issued a permanent registration tag, and no annual fee is required. New farm and residential tanks with a capacity less than 1,100 gallons also qualify for a permanent tag, even though they must comply with all other UST rules. Farm and residential tanks larger than 1,100 gallons must register annually and pay the annual fees.10/

A UST owner or operator who "intentionally" fails to comply with UST registration requirements faces a minimum penalty of $7,500.11/ Either the owner or operator of a UST can register it, but both are liable if both fail to do so.

Each time a UST is registered, the owner or operator must certify compliance with the "financial responsibility" or "financial assurance" laws. These laws require the owner or operator to demonstrate to the DNR that the owner or operator would have sufficient funds to pay for necessary cleanup and to compensate any injured third party for property damage or personal injury resulting from a release. This can be done through any one, or a combination of, the following methods: (1) insurance

---

9/ Iowa Code § 455B.473(7).

10/ Iowa Code § 455B.473.

11/ Iowa Code § 455B.477(6).

(public or private); (2) obtaining a guarantee, surety bond, letter of credit; (3) establishing a trust fund; or (4) qualification as a "self-insurer."12/ Unless a UST owner-operator buys and stores petroleum solely for its own use, an owner or operator must demonstrate minimum financial responsibility of $1 million per occurrence. If the UST owner owns more than 100 USTs, the owner or operator also must demonstrate aggregate coverage of at least $2 million.13/ Financial assurance requirements for local governments took effect February 18, 1994.

To show financial responsibility using a guarantee, surety bond, letter of credit, or trust fund, the owner or operator must use an instrument that conforms to the language in Iowa Admin. Code chapter 561-136. To show financial responsibility as a "self-insurer," a UST owner must demonstrate a tangible net worth exceeding $10 million and the ability to pass one of the financial tests described in 40 C.F.R. § 280.96.14/ For USTs that meet new UST performance standards or have already been upgraded, owners or operators can satisfy financial responsibility requirements through participation in a state-run insurance program financed in part by premiums paid by participants and in part by a tax on gasoline sales.

10.3 Performance Standards For New USTs

The UST rules contain performance standards for three areas: (1) corrosion protection; (2) leak detection; and (3) spill and overfill prevention. These performance standards are based on industry recommended practice guidelines, which are generally incorporated by reference in the rules. In some cases, separate performance standards are imposed for tanks and piping and for petroleum USTs and hazardous substance USTs.

12/ Iowa Code § 455B.474(2).

13/ Iowa Admin. Code § 561-136.4.

14/ Iowa Admin. Code § 561-136.4.

Strict standards apply to USTs installed after January 14, 1987 (called "new USTs"). Because the legislature intended to allow tank owners a period of time to "upgrade" or replace tanks installed before the adoption of the rules, tanks installed before January 14, 1987 (called "existing USTs") are not required to comply with all new UST performance standards now. Existing tanks must be equipped with leak detection now, however, and must be upgraded with corrosion protection and spill and overfill prevention equipment by December 22, 1998. The performance standards for upgraded USTs are somewhat different from the performance standards for new USTs due to special problems created by retrofitting an old tank with corrosion protection.

To meet the corrosion protection performance standards, a new UST must be: (1) constructed of fiberglass-reinforced plastic; (2) constructed of steel with an exterior coating and protected with a sacrificial anode or impressed current cathodic protection system; or (3) installed in an area demonstrated by a specialist not to be corrosive enough to cause leaks.[15] Impressed current cathodic protection systems must be inspected every sixty days; sacrificial anode systems must be inspected every three years.[16] All systems must be inspected and tested within six months of installation. To meet the spill and overfill prevention requirement, new USTs must be equipped with a device to prevent a release when the fill hose is detached from the fill pipe (typically a spill catchment basin) and a device to either stop flow before a tank is overfilled or emit an alarm when a UST is almost full.[17]

An owner or operator of a new UST system must also provide proof that the system was installed by a manufacturer-certified or state-licensed installer or inspected by a state-licensed engineer or UST inspector during installation.[18] The

---

15/ Iowa Admin. Code § 561-135.3(1).

16/ Iowa Admin. Code § 561-135.4(2).

17/ Iowa Admin. Code § 561-135.3(1)(c).

18/ Iowa Admin. Code § 561-135.3(1)(e).

installer must then sign the initial registration form certifying that the installation is in compliance with all applicable rules.

10.4 Performance Standards For Existing USTs

All existing USTs must be upgraded with corrosion detection and spill and overfill prevention, or replaced by December 22, 1998. Corrosion protection upgrading can be accomplished by installing impressed cathodic protection, installing an interior lining, or both. USTs that have been in the ground for more than 10 years can be upgraded only if they are first internally inspected, an expensive process which often involves some excavation, sandblasting the tank interior, and making repairs. For this reason, tanks older than ten years are often replaced rather than upgraded. Tanks less than ten years old can be upgraded after proving the integrity of the tank using existing monthly monitoring records or tank tightness testing.[19] When upgrading a UST, spill and overfill prevention equipment must also be installed.

10.5 Release Detection Requirements

All new and existing UST systems must be equipped with a method of release detection (phase-in dates for release detection have all passed). The release detection method must be capable of detecting a release from any portion of the tank and connected underground piping.[20] Owners and operators must install and operate the release detection method in accordance with the manufacturer's installation and maintenance requirements. Release detection methods must be capable of detecting a leak rate of 0.2 gallons per hour with a 95 percent probability of detection and a 5

19/ Iowa Admin. Code § 561-135.3(2).

20/ Iowa Admin. Code § 561-135.5(1).

percent probability of false alarm.[21/] When a release detection method indicates that a release may have occurred, owners and operators must notify the DNR.[22/]

For petroleum tanks with a capacity of 550 gallons or less, release detection requirements may be satisfied by using weekly tank gauging.[23/] For larger petroleum tanks that have not yet been upgraded, monthly inventory controls and annual tank tightness testing may be used until December 22, 1998.[24/] For petroleum tanks that have been upgraded and for new USTs, release detection procedures may combine monthly inventory control with tank tightness-testing every five years until December 22, 1998 or until 10 years after the upgrade or new installation, which ever is later.[25/]

Except for the above circumstances, all petroleum tanks must be monitored at least every thirty days using one of the following methods: automatic tank gauging, vapor monitoring, groundwater monitoring, interstitial monitoring for product between the tank and a secondary containment barrier (i.e., a double-walled tank), or other methods that may be approved by the DNR if the proposed method can detect a 0.2 gallon per hour leak rate or a cumulative monthly release of 150 gallons with a probability of detection of 95 percent and a probability of false alarm of 5 percent, or a method that the owner and operator can demonstrate is as effective as the approved methods.[26/] Technical requirements for each of these methods are listed at Iowa Administrative Code §§ 561-135.5(4)(d)-(h).

21/ Iowa Admin. Code § 561-135.5(1).

22/ Iowa Admin. Code § 561-135.5(1).

23/ Weekly tank gauging methods are specified in Iowa Admin. Code § 561-135.5(4)(b).

24/ Iowa Admin. Code § 561-135.5(2). After December 22, 1998, the tank must be upgraded or permanently closed.

25/ Iowa Admin. Code § 561-135.5(2).

26/ Iowa Admin. Code § 561-135.5(2).

If a tank is replaced at a site where a significant leak or spill has previously occurred, the replacement tank must be double-walled or be equipped with a secondary replacement system; a tank at such a site cannot be upgraded.[27/]

Petroleum UST piping must also have separate leak detection equipment. Pressurized piping systems must be equipped with an automatic line leak detector, and an annual line tightness test must be performed unless a leak detection method is used that is capable of detecting a line leak (such as double-walled piping with interstitial monitoring).[28/] Strict leak detection requirements are imposed for hazardous substance USTs, which generally require a secondary containment system with interstitial monitoring.[29/]

10.6 UST Closure or Change-in-Service Requirements

When a UST system is "temporarily closed," release detection may be stopped if the tank has been emptied, but corrosion protection must be continued unless the tank is empty. When a UST is temporarily closed for more than ninety days, vent lines must remain open and functioning, and all other lines must be capped and secured.[30/] A UST that has been upgraded or meets new UST performance standards can be temporarily closed indefinitely. An existing UST that has not been upgraded must be permanently closed after being temporarily closed for one-year. An extension of this one-year period can be obtained from the DNR if the UST owner first conducts a "site assessment."[31/]

Before closing a UST system permanently or changing its use to contain a non-regulated substance, certain testing requirements must be met. The DNR must be notified of the closure or change-in-service at least thirty days before beginning

---

27/ Iowa Admin. Code § 561-135.3(2)(a).

28/ Iowa Admin. Code § 561-135.5(2)(b).

29/ Iowa Admin. Code § 561-135.5(3).

30/ Iowa Admin. Code § 561-135.9(1).

31/ Iowa Admin. Code § 561-135.9(1).

work. To permanently close a UST, the system must be emptied of all liquids and accumulated sludges and tanks must be either removed from the ground or filled with an inert material.32/ Before completing either a permanent closure or change in service, a "site assessment" must be conducted, which entails testing the soil and groundwater for contamination in the areas where such contamination is most likely to be present.33/

10.7 Release Reporting Requirements

A UST owner must report both "suspected" and "confirmed" releases to the DNR within twenty-fours hours of discovery.34/ A "release" of petroleum product occurs whenever more than twenty-five gallons of product is spilled or whenever a smaller spill cannot be immediately cleaned up.35/ "Suspected releases" include any of the following situations:

1. Discovery of free product or vapors in soils, basements, sewer or utility lines, or nearby surface water;
2. Unusual and unexplained operating conditions of UST or dispensing equipment; or
3. Monitoring results from a release detection method that indicate that a release may have occurred.

In the case of inventory control, monitoring results give rise to a "suspected release" when results from two consecutive months indicate a loss greater than one percent

---

32/ Iowa Admin. Code § 561-135.9(2). The regulations define "empty" as containing less than one inch of sediment or less than 0.3 percent by weight of the total capacity of the UST system. Removing USTs often presents a problem because the sediment in the tank may contain a hazardous waste. Companies that pick up excavated tanks for disposal are required to notify the DNR that they are handling tank sediments, keep a record of the tank's original owners and what was stored in the tank, the volume of sediment, and what the salvager did with the sediment.

33/ Iowa Admin. Code § 561-135.9(3).

34/ Iowa Admin. Code § 561-135.6(1); see also chapter 8 of this handbook regarding release reporting requirements.

35/ Iowa Admin. Code § 561-135.6(4).

of flow-though plus 130 gallons.[36/] Inventory control results from a time period less than two months may give rise to a "suspected release" if they indicate a sudden and unusual product loss.

If a release is suspected, owners and operators must conduct a system test and a site check where contamination is most likely to be present.[37/] The system test requires owners and operators to test the system's tanks and piping, and to repair, replace, or upgrade the system if a leak exists.[38/] Even if no leak has been discovered from a particular tank, owners and operators may be required to conduct a site check if environmental contamination is discovered off-site to determine if the UST system is the source of any contamination.[39/] If the site check does not indicate that a release has occurred, no further investigation is required.[40/]

10.8 Release Response

Upon confirming a release, the UST owner or operator must report the release within 24 hours and take immediate action to prevent further release of the regulated substance.[41/] A trained groundwater professional should be immediately employed to supervise initial abatement measures. Within 20 days after reporting the release, the owner or operator must submit a report describing these initial abatement measures. An "initial site characterization report" must next be submitted describing the extent of contamination.[42/] Finally, the UST owner or

---

36/ Iowa Admin. Code § 561-135.5(4)(a) (incorporating by reference American Petroleum Institute Publication 1621, "Recommended Practice for Bulk Liquid Stock Control at Retail Outlets").

37/ Iowa Admin. Code § 561-135.8(1).

38/ Iowa Admin. Code § 561-135.8(1).

39/ Iowa Admin. Code § 561-135.8(1).

40/ Iowa Admin. Code § 561-135.8(1).

41/ Iowa Admin. Code § 561-135.7(1).

42/ Iowa Admin. Code § 561-135.7(4).

operator must undertake efforts to locate and remove any free product from the environment.[43]

After the completion of these initial steps, further release response must be completed as required by the DNR. The DNR may require completion of a detailed "site assessment report" describing the extent and risk of groundwater contamination.[44] Based on the level of risk of groundwater contamination created by the release, the DNR classifies every site as:

1. High risk;
2. Low risk; or
3. No-action required.[45]

At "high risk" sites, the owner or operator must evaluate corrective action design alternatives in a "site cleanup report" and implement an approved corrective action design plan.[46] A corrective action design can be approved by the DNR only after public comments have been heard and if the design utilizes "best available treatment technology" to remediate groundwater to achieve "maximum contaminant levels" or "action levels."[47] A list of conditions that, if present, would require the DNR to classify a site as "high risk" is codified at Iowa Admin. Code § 561-561-135.8(4).

At "low risk" sites, the owner or operator is required to implement "best management practices" and to conduct groundwater monitoring to assure the DNR that the release does not pose a risk of significant groundwater contamination. This monitoring may be required for up to twelve years.[48] The DNR will issue, if

---

43/ Iowa Admin. Code § 561-135.7(5).

44/ Iowa Admin. Code § 561-135.8(3).

45/ Iowa Code § 455B.474(1)(d).

46/ Iowa Admin. Code § 561-135.8(5).

47/ Iowa Admin. Code § 561-135.8(5).

48/ Iowa Admin. Code § 561-135.8(5).

requested by a UST owner, a "monitoring certificate" stating that no additional corrective action will be required unless contaminant levels increase.49/

At "no action required" sites, the owner or operator is not required to do any further corrective action or monitoring.50/ The DNR will issue a "certificate of completion of site remediation" under the following circumstances: (1) a site is classified as a "no action required;" (2) the monitoring period has ended at a "low risk" site; or (3) the "site cleanup report" has been successfully implemented at a "high risk" site.

10.9 Enforcement

The DNR has broad enforcement authority to enforce the UST laws. Any person who violates any of the UST rules may be ordered to perform corrective action and subject to a civil fine of $5,000 per day.51/ The DNR also has authority to issue administrative penalties of not more than $10,000, after an opportunity for a contested case hearing.52/ Persons who knowingly misrepresent the condition of a UST, fail to register a tank, or fail to report a leak may be guilty of an aggravated misdemeanor.53/ Under Iowa's citizen suit statute, any person adversely affected may commence a civil action against any person for violating Iowa's storage tank statutes and rules.54/

The DNR interprets Iowa statutes relating to water quality and hazardous conditions as authorizing the DNR to order site assessment and cleanup.55/ If a UST

49/ Iowa Admin. Code § 561-135.8(9).

50/ Iowa Code § 455B.474(1)(d); Iowa Admin. Code § 561-135.8(4).

51/ Iowa Code § 455B.476, .477.

52/ Iowa Code § 455B.109.

53/ Iowa Code § 455B.477.

54/ Iowa Code § 455B.111.

55/ In re Dell Oil Co., DIA No. 92DNR-1, 1992 Westlaw 348,892 (Sept. 30, 1992), adopted, 1993 Westlaw 475,977 (Nov. 5, 1993) (relying on Iowa Code §§ 455B.175, 455B.186, and 455B.388); In re Clapp, DIA No. 89DNR-34, 1989 Westlaw 198,596 (Nov. 9, 1989) (relying on Iowa Code §§

owner or operator fails to respond to a release as required by the rules, and the DNR chooses to undertake corrective action on its own, the UST owner and operator will be held strictly liable in a cost recovery action by the DNR and the UST Board. In addition, the owner or operator may be liable for punitive damages equal to three times the actual cost of the corrective action.56/

10.10 Compensation For Cleanup Costs

Iowa has been a leader among states using public funds to compensate petroleum UST owners for taking corrective action and to achieve early compliance with UST rules. The heart of the Iowa Financial Responsibility Program -- originally intended as a means to enable owners to demonstrate financial responsibility as required by the federal UST rules -- is the Iowa Comprehensive Petroleum Underground Storage Tank Trust Fund (UST Fund). It is funded by a use-tax on the sale of gasoline, 73 percent of the annual collected storage tank management fees, premiums collected by the state-run insurance program, penalties paid for violation of the USTs rules, and the proceeds of any corrective action cost recovery action.57/ This fund is then divided between three independent programs: (1) the remedial program; (2) the loan guarantee program; and (3) the insurance program. These programs, although supervised by a "UST Board" and the DNR, are currently implemented by a private contractor, Williams & Company.

10.10.1 UST Remedial Program and Community Remediation Program

The remedial program was originally intended to encourage UST owners to upgrade or replace old USTs and to report and remediate past releases as soon as possible. The program provided direct compensation for corrective action costs

455B.186, 455B.381).

56/ Iowa Code § 455G.13.

57/ Iowa Code §§ 455B.476, 455B.477. The UST Fund rules are codified at Iowa Admin. Code ch. 591-1 to 17.

incurred by an owner or operator for releases reported to the DNR between 1985 and 1990. Although the eligibility dates were subsequently extended, and the fund continues to provide compensation for claims at sites where corrective action is ongoing, remedial fund compensation is unavailable for releases reported today. Unless a legislative change is made, in the future the remedial account can be used only to fund corrective action at sites where a solvent owner or operator cannot be found.58/

Although the remedial program is unavailable for future releases, a supplemental program created through it may continue to assist UST owners and operators. Under the Community Remediation Program, allowed by 1991 legislation, the UST Board can designate a city as a "Community Remediation Project" (CRP) if there are at least two sites eligible for remedial program funding in the city.59/ The UST Board can the solicit bids from environmental consultants to complete all of the leak investigation and corrective action necessary at all of the sites within the CRP. By selecting one contractor to do all the work in a city, the UST Board has been able to reduce costs. The UST Board now allows parties who are not eligible for remedial program funding to "buy in" to the CRP at the average cost per site. In many cases, this may create a substantial savings. Because the UST Board intends to have ninety cities within the community remediation program by December of 1994, this program may be a substantial benefit to UST owners and operators.60/

10.10.2 UST Loan Guarantee Program

The loan guarantee program is intended to enable small businesses to receive the financing they need to upgrade their UST systems and conduct any required corrective action. To qualify for a loan guarantee, a UST owner must be an

---

58/ Iowa Code § 455G.9.

59/ Iowa Code § 455G.2(5); Iowa Admin. Code § 591-13.2.

60/ DNR Tanks Update, v. 3, #1, p. 3 (April 1994).

independently owned and operated business, own less than twelve tanks at no more than two locations, and have a net worth of $400,000 or less.61/ In addition, the UST owner must demonstrate an inability to obtain private financing without the loan guarantee.62/

10.10.3 UST Insurance Program

The insurance program will be the most important UST Fund program in the future. The insurance program is intended to enable UST owners to obtain insurance, which may not be otherwise available, to meet federal financial responsibility requirements. Coverage is limited to the levels of financial responsibility required by federal law, $1 million per incident and $2 million aggregate for a petroleum marketer with more than 100 tanks.63/

The eligibility provisions of the insurance account statute are not clearly drafted. They have been interpreted by the UST Board to allow USTs be covered at the basic premium rate if they have been upgraded or meet the new UST performance standards. Existing USTs that have not been upgraded may be covered under the program at a higher premium rate until January 1, 1995. After that date, existing USTs that have not been upgraded will be ineligible. In addition, because the insurance program is intended to protect UST owners and operators in the event of future releases, and not to pay for remediation of past releases, full coverage is only available for sites that have been tested and found to be "release free" as of the date that coverage began. At present, however, the UST Board will allow UST owners to obtain coverage at sites where remediation is on-going. This coverage is limited to liability arising out of a future release. In the event of a future release, the burden will be on the insured to distinguish liability incurred as a result of the future release from liability incurred as a result of a past release. At the time

61/ Iowa Code § 455G.2(18); Iowa Admin. Code § 591-12.2.

62/ Iowa Code § 455G.10(5); Iowa Admin. Code § 591-12.1(15)(b).

63/ Iowa Code § 455G.11(2).

of this writing, there have not yet been reported cases addressing this difficult causation issue.

The current premium rate for the insurance program is $350 per tank ($750 for existing tanks that have not been upgraded). As of August of 1994, the UST Board is required to annually adjust premium levels on an actuarially sound basis.64/ Williams and Company is currently in the process of performing the required actuarial analysis. It is likely that premiums will increase significantly at the first annual adjustment.

Insurance is also available for UST installers 65/ and for the transfer of property for which a monitoring certificate or a certificate of completion of site remediation has been issued.66/

10.11 Recordkeeping

The UST rules contain extensive requirements to preserve installation records and manufacturer warranties (life of UST), repair records (life of UST), release detection records (1 year), corrosion protection inspection records (generally the last 3 inspection reports), closure records (3 years), and release reporting records (generally 1 year, but tank tightness test must be saved until next test is complete). These records must be maintained at the UST site or at an alternative site where they are readily available for inspection upon demand by the DNR.67/ Owners or operators may find it useful to maintain the records longer than the minimum periods specified above in the event problems or claims arise.

64/ Iowa Code § 455G.11(2).

65/ Iowa Code § 455G.11(6).

66/ Iowa Code § 455G.11(10); Iowa Admin. Code § 591-10.5.

67/ Iowa Admin. Code. R. 561-135.4(5).

10.12 Property Transfer -- Recording of USTs

Documents relating to title to real property cannot be recorded by a county recorder in Iowa unless the seller submits a signed declaration describing the size, type, and contents of any USTs on the property or stating that there are no known USTs on the property.[68]

10.13 Fire Code Regulations

The Iowa State Fire Marshall has adopted the National Fire Protection Association Automobile and Marine Service Code (NFPA 30A), which also regulates spill prevention and leak detection from USTs.[69] The fire code rules are generally no more strict than the DNR UST rules described above with regard to corrosion protection, leak detection, and spill and overfill prevention. However, the fire code contains many regulations relating to petroleum USTs that are not also codified in the DNR rules. These include, for example, regulation of dispensing equipment, signage, and fill pipes.

10.14 Aboveground Storage Tank (AST) Regulations

In 1989, the Iowa legislature passed a statute requiring registration with the State Fire Marshall of aboveground petroleum storage tanks with a capacity greater than 1,100 gallons.[70] The registration requirement does not apply to heating oil tanks used for consumptive use on the premises. This statute requires registration of tanks only. It does not create performance standards or reporting and corrective action requirements. However, ASTs cannot be installed or modified without plan approval from the State Fire Marshall.[71]

The State Fire Marshall has adopted NFPA 30A and the Flammable And

---

68/ Iowa Code § 558.69.

69/ Iowa Admin. R. 661-5.304(1).

70/ Iowa Code § 101A.21.

71/ Iowa Admin. Code § 661-5.301(1).

Combustible Liquids Code (NFPA 30).72/ These codes contain requirements relating to the location and capacity of ASTs, control of spillage, and corrosion protection. With regard to petroleum storage facilities with ASTs larger than 40,000 gallons, Iowa has adopted the federal new source performance standards for air emissions.73/

10.15 Spill Prevention, Control and Countermeasure Plans

In addition to the state regulations discussed above, ASTs and certain USTs may be subject to oil pollution prevention regulations adopted pursuant to the federal Clean Water Act and the Oil Pollution Act. If a non-transportation-related oil storage facility is located such that if a release were to occur, it "could reasonably be expected" to contaminate navigable waters, then the facility owner or operator is required to create a Spill Prevention, Control and Countermeasure (SPCC) plan.74/ EPA regulations exempt the following facilities from SPCC plan requirements:

1. A facility with an aggregate AST capacity less than 1,320 gallons, provided that no single AST is larger than 660 gallons; and

2. A facility with a total UST capacity of 42,000 gallons or less.75/

Accordingly, facilities with oil storage capacities larger than these quantities generally are presumed to require SPCC plans.

An SPCC Plan is a comprehensive contingency plan, approved by a certified engineer, which must explain the facility's compliance with relevant regulations and set forth the measures taken to prevent a release as well as the steps that will be taken in the event of a release.76/ AST owners should be aware that the SPCC

72/ Iowa Admin. Code §§ 661-5.300, 661-5.304(1).

73/ Iowa Admin. Code § 567-23.1(2) (incorporating by reference 40 C.F.R. §§ 60.110-60.115a).

74/ 40 C.F.R. § 112.1.

75/ 40 C.F.R. § 112.1(2).

76/ 40 C.F.R. § 112.7.

regulations are currently under review by the EPA and may be amended to impose stricter regulation of ASTs.

10.15.1 Oil Pollution Act Facility Response Plans

In July 1994, EPA published final regulations regarding "facility response plans" that add supplemental emergency planning requirements for certain facilities.77/ For non-transportation-related onshore facilities, response plan requirements apply to:

1. Facilities that have a total oil storage capacity greater than 42,000 gallons which transfer oil over water to or from vessels; and
2. Certain other facilities with oil storage capacity greater than 1 million gallons.

Effective August 30, 1994, facilities that prepare an SPCC plan must analyze the applicability of the facility response plan regulations and, if the response plan regulations are not applicable, complete an EPA form to be kept with the SPCC plan verifying that a response plan is not required.78/

77/ 59 Fed. Reg. 34,070 (July 1, 1994) (to be codified as part of the SPCC regulations at 40 C.F.R. § 112.20).

78/ 40 C.F.R. § 112.20(e). Similar process safety management planning requirements apply to processes involving hazardous chemicals under OSHA regulations at 29 C.F.R. § 1910.119 and Clean Air Act regulations under consideration at 59 Fed. Reg. 4478 (Jan. 31, 1994).

CHAPTER ELEVEN

AGRICULTURAL CHEMICALS

11.0 Introduction To Agricultural Chemicals Regulation

The Iowa Department of Agriculture and Land Stewardship (DALS) is responsible for the regulation of pesticides, fertilizers, and certain other agricultural chemicals under Title X of the Iowa Code.[1/] Various Iowa Code chapters regulate fertilizers and soil conditioners, pesticides, chemigation practices, and agricultural lime materials.[2/] Each of these chapters is discussed below. The DALS's general regulatory powers are set forth in Iowa Code chapter 189.

Iowa Code chapter 189 gives the DALS broad authority over "articles," a statutory term including, among other things, commercial fertilizers, insecticides, and fungicides.[3/] The DALS is authorized to enter property, inspect facilities, and procure and analyze samples.[4/] Chapter 189 also requires that all articles in a package must be labeled with, among other information, brand name, quantity of contents, and name and location of the manufacturer, importer, distributor or dealer.[5/] In addition, unless otherwise provided by law, articles which are mixtures, such as fertilizer-pesticide mixtures, must be marked as such and must identify the names of all ingredients contained therein.[6/] Chapter 189 specifically prohibits any label representations which are deceptive as to the true character of an article or the

---

1/ Iowa Code ch. 189-216.

2/ For further discussion of agricultural issues, see also sections 3.3 (animal feeding permits), 3.9 (well closing requirements), 4.4.4 (farmed wetlands), and 14.1.1 (farmer protections from nuisance claims) of this handbook.

3/ Iowa Code § 189.1(1). The term "commercial fertilizer" includes fertilizers, soil conditioners, and fertilizer-pesticide mixtures. The term "article" as defined in Iowa Code § 189.1 does not specifically include pesticides, but the "article" definition does include insecticides and fungicides, which fall within the definition of "pesticides" in Iowa Code ch. 206. Thus, although not free from ambiguity, it appears that the general provisions of Iowa Code ch. 189 do apply to pesticides.

4/ Iowa Code §§ 189.3-189.7.

5/ Iowa Code § 189.9.

6/ Iowa Code § 189.11.

place of its production.[7] It also is illegal for any person to knowingly introduce into Iowa, solicit orders for, deliver, transport, or have in his or her possession with intent to sell any article which is labeled in violation of Iowa law.[8] Chapter 189 also provides that no person shall knowingly manufacture, introduce into the state, solicit orders for, sell, deliver, transport, or offer for sale in any fashion any article which is adulterated.[9]

The DALS is authorized to confiscate or condemn any article that is adulterated or improperly labeled.[10] DALS also has authority to obtain injunctive relief restraining any person engaging from any business for which a license is required without obtaining such a license.[11] In addition, any person violating any provision of Title X of the Iowa Code, or any regulation enacted by the DALS thereunder, is guilty of a simple misdemeanor for each offense committed.[12] These general enforcement authorities are in addition to those more specific provisions discussed below.

11.1 Fertilizers and Soil Conditioners Regulation

The DALS regulates fertilizers and soil conditioners under the Iowa Fertilizer Law.[13] The term "fertilizer" generally includes any substance containing one or more recognized plant nutrients designed to promote plant growth.[14] A "soil conditioner" is any substance intended to provide a favorable growth, yield, or

---

7/ Iowa Code § 189.13.

8/ Iowa Code § 189.14.

9/ Iowa Code § 189.15.

10/ Iowa Code § 189.17.

11/ Iowa Code § 189.20.

12/ Iowa Code § 189.21.

13/ Iowa Code ch. 200.

14/ Iowa Code § 200.3(9). Fertilizer does not include animal or vegetable manures or calcium and magnesium carbonate materials used primarily for correcting soil acidity.

quality of crop, soil, flora or fauna, or other soil characteristics, but which is not a fertilizer, agricultural liming material, or pesticide.[15/]

Under chapter 200, the DALS has authority to exercise oversight over various aspects of the distribution and use of fertilizers and soil conditioners. The DALS Secretary has utilized this authority to adopt numerous regulations.[16/]

11.1.1 Registration and License Requirements

No person may sell, offer for sale, distribute, manufacture, mix, custom mix to order, or blend any fertilizer or soil conditioner without obtaining a license from the DALS for each location from which fertilizer or soil conditioner is manufactured or distributed within the state.[17/] With limited exceptions, each brand and grade of fertilizer and soil conditioner must be registered with the DALS before it is sold, offered for sale, or distributed in the state.[18/]

The basic registration information required -- including product name and brand, grade, and guaranteed analysis -- is set out in Iowa Code § 200.5. The DALS Secretary also may require a person applying for a registration to sell or distribute any fertilizer or soil conditioner in Iowa to submit additional data to substantiate claims made for the product.[19/] The Secretary must give notice if the information submitted is deficient so that a registration applicant may have an opportunity to make necessary corrections.[20/]

A person also must be licensed with the DALS to mix pesticides and fertilizers, and such mixtures must be labeled and registered in accordance with both

---

15/ Iowa Code § 200.3(22).

16/ See generally Iowa Admin. Code ch. 21.

17/ Iowa Code § 200.4(1).

18/ Iowa Code § 200.5(1). Fertilizers which are custom formulated to order are exempt from certain registration requirements. Id.

19/ Iowa Code § 200.5(6).

20/ Iowa Code § 200.5(7).

the Iowa pesticide and fertilizer laws.21/ The mixture licensee must at all times produce a uniform mixture of fertilizer or soil conditioners and pesticide.22/

The DALS may review fertilizer and soil conditioner registrations to determine if any product claims made are accurate.23/ The rules establish standards for a producer or distributor to substantiate product claims.24/ If the DALS determines that a product does not meet the claims by the producer or distributor or the minimum nutrient values established by the state, the DALS may, among other enforcement powers:

1. Order the product deficiencies to be corrected;
2. Require purchasers of the product to be reimbursed; and
3. Revoke, suspend or refuse to grant any registration or license necessary under chapter 200 to distribute the product.25/

11.1.2 Fertilizer Regulation

The DALS has enacted rules to prevent the use and storage of fertilizers and soil conditioners in ways which may endanger humans, damage agricultural products, or cause unreasonable adverse affects on the environment. For example, the DALS has promulgated detailed rules regulating many of the operational aspects of storing and handling anhydrous ammonia.26/ The DALS also regulates the application of fertilizers through irrigation systems, and any person operating such a system must obtain a chemigation permit.27/

21/ Iowa Code 200.4(1); see also section 11.2 of this handbook regarding the Iowa Pesticide Act.

22/ Iowa Code § 200.4(2).

23/ Iowa Admin. Code § 21-43.23.

24/ Iowa Admin. Code § 21-43.24.

25/ Iowa Admin. Code § 21-43.5.

26/ Iowa Admin. Code § 21-43.6.

27/ Iowa Code ch. 206A; see also section 11.3 of this handbook regarding chemigation restrictions.

DALS rules also require facilities storing fertilizers and soil conditioners, both liquid and dry, to be maintained and operated in a manner that prevents contamination, and specific safeguards are required by the DALS to prevent, among other things, spills of liquid fertilizers and particulate emissions from dry fertilizers during loading, unloading, and mixing of product, and washing of equipment.28/ By February 18, 1997, all such storage facilities subject to these DALS rules must meet various secondary containment requirements.29/ Design plans and specifications for any such facility must be submitted to the DALS before construction is commenced.30/ After the facility is complete, a certification must be submitted to the DALS confirming construction in accordance with the Department's requirements.31/

11.1.3 Labeling Requirements

It is illegal to sell or distribute fertilizers or soil conditioners in bags or containers without labels identifying, among other information, brand name and grade, guaranteed analysis, and registrant name and address.32/ There are additional labeling requirements applicable to fertilizers that are formulated to a customer's order and to fertilizers distributed in bulk.33/

A person may not sell or distribute any misbranded fertilizer or soil conditioner. Misbranding occurs when a fertilizer does not identify substances promoting plant growth, or when false or misleading statements are included on

---

28/ Iowa Admin. Code §§ 21-44.50-.58. Nonliquid fertilizers stored in totally enclosed facilities generally are exempt from these rules. Iowa Admin. Code § 21-44.54.

29/ Iowa Admin. Code § 21-44.50. Beginning in February 1987, all new storage facilities were required to be constructed to meet these secondary containment requirements.

30/ Iowa Admin. Code § 21-44.52.

31/ Iowa Admin. Code § 21-44.52.

32/ Iowa Code §§ 200.5(1), (6).

33/ Iowa Code §§ 200.5(2), (3), (4).

the product container or label or are disseminated in any other manner.[34] It also is unlawful for any person to manufacture, offer for sale, or sell in Iowa any commercial fertilizer or soil conditioner containing any fill that is injurious to crop growth or deleterious to soil, or to use as filler any inert or useless plant food material for the purpose of deceiving or defrauding the purchaser.[35]

11.1.4 Enforcement

The DALS Secretary may cancel the registration of any commercial fertilizer[36] or soil conditioner or the license of any licensee of a commercial fertilizer or soil conditioner, or may refuse to grant such registration or license, if there is evidence of any fraudulent or deceptive practices or willful violation of the Iowa Fertilizer Law or DALS rules by the registrant or licensee.[37] This cancellation authority is in addition to the Secretary's suspension and revocation powers with respect to fertilizers and soil conditioners not meeting product claims or minimum nutrient value requirements discussed in section 11.1.1 of this handbook If there is any violation of the law with respect to any commercial fertilizer or soil conditioner, the Secretary may issue "stop sale" orders or seize and dispose of the product.[38] The Secretary generally must provide notice and opportunity to cure defects before exercising these remedies. The DALS Secretary also may prosecute any violation of the Fertilizer Law as a simple misdemeanor or may seek injunctive relief to require compliance with the law.[39]

34/ Iowa Code § 200.12.

35/ Iowa Code § 200.11.

36/ A "commercial fertilizer" is any fertilizer, fertilizer material, or fertilizer pesticide mixture. Iowa Code § 200.3(5).

37/ Iowa Code § 200.15.

38/ Iowa Code §§ 200.16, 200.17.

39/ Iowa Code § 200.18.

11.2 Pesticide Control And FIFRA

The Iowa Pesticide Act is codified in Iowa Code chapter 206. The term "pesticide" is defined as any substance or mixture of substances:

1. Intended to prevent, destroy, repel, or mitigate, directly or indirectly, any insects, rodents, rematodes, fungi, weeds and other forms of plant or animal life or viruses; or

2. Intended for use as a plant growth regulator, defoliant, or desiccant.[40/]

Pesticide regulations are generally administered by the DALS. Iowa law incorporates by reference certain restrictions contained in the Federal Insecticide, Fungicide and Rodenticide Act (FIFRA).[41/] FIFRA creates a complex and comprehensive federal regulatory program for pesticides that exists in addition to the state laws and regulations outlined in this handbook.

11.2.1 Registration Requirements

All pesticides distributed, sold or offered for sale for use in Iowa or delivered for transportation or transported in intrastate commerce between points within the state through any point outside the state must be registered with the DALS.[42/] The information required for registration, provisions for confidential treatment of trade secrets, and registration fee amounts generally are established in Iowa Code § 206.12. Pesticide registrations expire on December 31 of each year and may be renewed on or before that date for the following year. Registration is not required if a pesticide is shipped from one plant within Iowa to another within the state operated by the same person.[43/]

The DALS Secretary is required to determine, by rule, the pesticides to be

---

40/ Iowa Code § 206.2(20).

41/ 7 U.S.C. §§ 136-136y; 40 C.F.R. pts. 152-186.

42/ Iowa Code §§ 206.11, 206.12(1).

43/ Iowa Code § 206.12(6).

classified and registered as restricted use pesticides.[44] The Secretary has adopted restricted use classifications enacted by the United States Environmental Protection Agency (EPA) to satisfy this requirement.[45]

The DALS Secretary may approve special local need registration where no pesticide registered with the EPA is available for the necessary use or where an EPA-registered product is not available or would not be safe or efficacious.[46] The DALS has adopted by reference certain FIFRA requirements concerning special local need registration, and has authority to revoke, suspend, or deny such registration if certain adverse effects may result from use of the product.[47]

11.2.2 License and Certification Requirements

The Iowa Pesticide Act establishes a number of license and certification requirements regarding the use of pesticides. Persons who sell or offer for sale restricted use pesticides, pesticides for use by commercial or public applicators, or general use pesticides labeled for agriculture or lawn and garden use must obtain a pesticide dealer license from the DALS for each location from which distribution occurs.[48] Manufacturers, registrants, or distributors located outside of Iowa that have no in-state distribution outlet and distribute pesticides in the state also must have an Iowa pesticide dealer license.[49] Applicators who apply pesticides owned and supplied by others need not obtain a dealer license if they do not charge for the sale of the pesticides, but they must have an applicator's license as discussed

---

44/ Iowa Code § 206.20.

45/ Iowa Admin. Code § 21-45.30.

46/ Iowa Admin. Code § 21-45.6.

47/ Iowa Admin. Code § 21-45.6.

48/ Iowa Code §§ 206.2(21), 206.8(1). A pesticide dealer license need not be obtained by a person selling less than $10,000 of general use pesticides from each of his or her business locations. Iowa Code § 206.2(21).

49/ Iowa Code § 206.8(1).

below.[50/] Governmental agencies using pesticides in their own programs also are exempt from the dealer license requirement.[51/] All dealer licensees must file an annual report with the DALS.[52/]

Commercial and public applicators of pesticides must be licensed by the DALS and must comply with education and certification requirements established by the DALS before applying any pesticides to property of others.[53/] Commercial applicators also must provide satisfactory proof of financial responsibility, generally in the form of a surety bond or liability insurance, to the DALS.[54/] No person -- including commercial, public, structural, or private applicators -- may apply restricted use pesticides without first complying with the DALS's license, certification, and education requirements concerning such pesticides.[55/] Applicators who apply pesticides by air and structural pest control applicators must fulfill additional certification and education requirements.[56/] Commercial, public, and private applicators also are required to comply with chemigation certification and education requirements before applying pesticides through irrigation systems.[57/]

The time periods for which initial licensing, certification, and education requirements remain valid vary with the type of application activities involved, as do license renewal, recertification, and continuing education requirements.[58/] There are limited exceptions to these various requirements for persons working

---

50/ Iowa Code § 206.8(3)(a).

51/ Iowa Code § 206.8(3)(b).

52/ Iowa Code § 206.12(7).

53/ Iowa Code §§ 206.5, 206.6, 206.7, 206.31; Iowa Admin. Code § 21-45.22.

54/ Iowa Code § 200.13.

55/ Iowa Code §§ 206.5, 206.31; Iowa Code § 206.5; Iowa Admin. Code § 21-45.22.

56/ Iowa Code §§ 206.6(1);206.31.

57/ Iowa Code § 206.5(6).

58/ Iowa Code §§ 206.5, 206.6, 206.7; Iowa Admin. Code §§ 21-45.22, 45.52.

directly under the supervision of certified applicators, farmers applying pesticide on their own land, and veterinarians.59/ These exceptions, however, are limited and must be reviewed carefully to determine their applicability before being relied on by any person.

The DALS requires commercial applicators to maintain records with respect to their applications of pesticides.60/ Each commercial applicator must keep those records for three years after the date of application, and must make the records available to the DALS upon request.61/

11.2.3 Labeling Requirements

All pesticides distributed in Iowa must be labeled in accordance with the requirements of the Iowa Pesticide Act. Required label information includes the following:

1. Name and address of the manufacturer, registrant, or person for whom the pesticide is manufactured;
2. Name, brand, or trademark;
3. Net weight or measure of the contents;
4. An ingredient statement; and
5. Date of manufacture.62/

Any pesticide containing any substance in quantities highly toxic to humans must include additional labeling information, including a skull and cross-bones mark, the word "poison" prominently displayed, a statement of the antidote, and instructions for safe disposal of the container.63/ Certain highly toxic pesticides identified under

---

59/ Iowa Code §§ 206.5, 206.18(3), (5).

60/ Iowa Admin. Code §21-45.26.

61/ Iowa Code § 206.15.

62/ Iowa Code § 206.11(1)(d); see also Iowa Admin. Code §§ 21-45.7-.14.

63/ Iowa Code § 206.11(1)(e). The standard for determining whether a pesticide is highly toxic is generally set forth at Iowa Admin. Code § 21-45.21.

chapter 206 must be distinctly colored or discolored in accordance with DALS regulations rather than distributed as a white powder.64/ It is unlawful for any person to detach, alter, deface, or destroy any label, or to add any substance to, or take any substance from, the pesticide in a manner that would interfere with labeling requirements.65/

It also is unlawful for any person to distribute, give, sell, or offer for sale within Iowa, or to deliver for transportation or transport in intrastate commerce or between points within the state through any point outside of the state, any of the following:

1. Any pesticide that differs in substance from the representations on the label or otherwise made concerning the pesticide with regard to its use;

2. Any pesticide that differs from the composition represented in connection with the registration, unless the DALS authorizes a change in labeling or formula; or

3. Any pesticide that is adulterated, misbranded or not in an unbroken container or does not have a label affixed to the container containing all information required by law.66/

11.2.4 Pesticide Use Regulation

The Pesticide Act requires generally that all pesticides must be handled, transported, stored, displayed, and distributed in a manner such that they do not endanger human beings or the environment, and do not endanger food, feed, or other products that may be transported, stored, displayed, or distributed with the pesticides.67/ It is unlawful for any person to:

1. Use or cause to be used any pesticide contrary to its labeling or applicable DALS rules; or

---

64/ Iowa Code § 206.11(1)(f); Iowa Admin. Code § 21-45.15.

65/ Iowa Code § 206.11(2)(a).

66/ Iowa Code § 206.11(1).

67/ Iowa Code § 206.11(3)(c).

2. Dispose of, discard, or store pesticides or pesticide containers in such a manner to cause injury to humans, vegetation, crops, livestock, wildlife, pollinating insects, or as to pollute any water supply or waterway.[68]

The DALS is required to promulgate standards concerning disposal of pesticide containers.[69] Distribution of any restricted use pesticide is prohibited to any person who is required by law to be certified to use or purchase restricted use pesticides unless that person (or the person's agent) is in fact certified to use or purchase such pesticides.[70]

Under the Pesticide Act, pesticide applicators may not take any of the following actions:

1. Make any pesticide recommendation or application inconsistent with labeling information;
2. Apply known ineffective or improper materials;
3. Operate faulty or unsafe equipment;
4. Operate in a faulty, careless or negligent manner; or
5. Fail to comply with limitations or restrictions provided in any license, permit or certification or any rule or order of the DALS.[71]

It is unlawful for applicators to fail to keep and maintain records required by law, or to make false or fraudulent records or reports.[72] Applicators also are specifically prohibited from aiding or abetting other persons in evading the requirements of chapter 206 or from making false or misleading statements during or after an inspection concerning pesticides.[73]

68/ Iowa Code §§ 206.11(3)(a), (d).

69/ Iowa Code § 206.25.

70/ Iowa Code § 206.11(3)(a).

71/ Iowa Code §§ 206.11(4)(a)-(e).

72/ Iowa Code §§ 206.11(4)(f), (g).

73/ Iowa Code §§ 206.11(4)(i), (j).

11.2.5 Bulk Pesticide Storage

The DALS has adopted rules for bulk pesticide storage to minimize potential adverse affects on the environment.74/ All nonmobile bulk pesticide storage containers must be located within secondary containment facilities meeting design standards specified by the DALS.75/ Design plans and specifications must be submitted to the DALS before facility construction is commenced.76/ After completion, a certification must be submitted to the DALS confirming that the facility was constructed in accordance with the Department's rules.77/

The rules also regulate various operational matters. Mixing, repackaging, and transferring of pesticides at bulk facilities, and washing of equipment in connection with such operations, generally must occur within containment facilities, and must meet other specified requirements to prevent adverse effects to humans or the environment.78/ If there is a discharge *within* a secondary containment facility, the operator must promptly recover the release to the maximum extent possible.79/ If there is a discharge *from* the facility, the operator must notify DNR as soon as possible (but in no event later than six hours) and recover the spill to the maximum extent possible.80/

74/ Iowa Admin. Code §§ 21-44.2-.10. A bulk storage container is one in which pesticides are transported or held in an individual container of greater than 55 U.S. gallon liquid measure or 100 pounds net dry weight.

75/ Iowa Admin. Code §§ 21-44.2-.3.

76/ Iowa Admin. Code § 21-44.3.

77/ Iowa Admin. Code § 21-44.4.

78/ Iowa Admin. Code §§ 21-44.2-.10.

79/ Iowa Admin. Code § 21-44.7(4).

80/ Iowa Admin. Code § 21-44.7(6); see also chapter 8 of this handbook regarding release reporting requirements.

11.2.6 Urban Use Restrictions

The Pesticide Act authorizes the DALS to enact requirements to control pesticide application within cities. Under this authority, the DALS has established limited urban pesticide application rules.[81] Notification signs must be posted at the start of an application and must remain in place for 24 hours. There also are a variety of specific application requirements that vary depending on the type (e.g., playground, golf course) and location of the area where the application is occurring.[82]

11.2.7 Enforcement

The Pesticide Act establishes several enforcement tools for the DALS to utilize. The Act authorizes confiscation and disposal of any pesticide that is adulterated or misbranded, has not been registered, fails to bear required labeling information, or is a white powder pesticide that has not been colored or discolored as required by law.[83] The judicial procedures applicable to this confiscation remedy are set forth in the statute.[84] When the DALS has reasonable cause to believe that a pesticide is being distributed, stored, transported or used in violation of the law, the Secretary may issue a written "stop sale, use or removal" order upon the owner or custodian of any such pesticide. The pesticide may not then be sold, used or removed until the provisions of chapter 206 have been complied with and the pesticide has been released in writing under conditions specified by the Secretary.[85]

Any person violating the registration requirements of Iowa Code § 206.11(1)(a) is guilty of a simple misdemeanor. Any person violating other

---

81/ Iowa Ann. Code § 21-45.50.

82/ Iowa Ann. Code § 21-45.50.

83/ Iowa Code § 206.16(1)(a). The Act does not expressly limit use of this confiscation authority to the IDA or other state entities.

84/ Iowa Code §§ 206.16(2), (3).

85/ Iowa Code § 206.16(4).

provisions of chapter 206 is guilty of a serious misdemeanor.86/ Any violation of chapter 206, including provisions relating to registration, committed with intent to defraud is a serious misdemeanor.87/ Certain violations also require the immediate termination of pesticide registrations.88/ The DALS is given authority to pursue these sanctions.

The DALS also may seek civil penalties on commercial applicators in an amount of $500 per day per offense.89/ A peer review procedure for alleged commercial applicator violations has been established under the Pesticide Act.90/ Either the applicator or the DALS may request review by the peer review panel (consisting of persons engaged in various applicator activities) within 14 days following DALS notice of a potential violation and proposed penalty.91/ The panel may recommend that the proposed penalty be increased, decreased, or eliminated, and that conditions be placed upon the applicator's license.92/ If the parties do not agree to accept the recommendation, then the DALS can proceed with a contested case.93/

The Pesticide Act also requires reporting of significant pesticide accidents or incidents to the DALS. Any person claiming damages from a pesticide application is required to file a claim with the DALS (on forms provided by the Department) within 60 days after the alleged damages occurred.94/ If a growing crop is alleged to

---

86/ Iowa Code § 206.22(2).

87/ Iowa Code § 206.22(3).

88/ Iowa Code § 206.22(2).

89/ Iowa Ann. Code § 21-45.102.

90/ Iowa Code § 206.23A.

91/ Iowa Ann. Code § 21-45.103.

92/ Iowa Ann. Code § 21-45.104.

93/ Iowa Ann. Code § 21-45.105.

94/ Iowa Code § 206.14(2).

have been damaged, however, the report must be filed prior to the time that 25 per cent of the crop has been harvested. The DALS must inspect damages claimed whenever possible, and when it determines that a complaint has merit, the DALS must make such information available to the person claiming damage and the person who is alleged to have caused the damage.95/ Where damage is alleged to have occurred, the claimant must permit the DALS, as well as the licensee and its representatives, to observe the damage.96/ Failure to permit such observation will bar the claim against the licensee.

11.3 Chemigation Regulation

Under Iowa Code chapter 206A, fertilizers, pesticides, and plant growth regulators may not be applied to land or plants through irrigation distribution systems without a chemigation permit obtained from the DALS.97/ The permit must be issued to the title holder of the land or the person responsible for day-to-day management of the chemigation system. The permit must be renewed every year under procedures established by the DALS.98/ The DALS may disapprove a permit application for good cause.99/

The DALS is responsible for establishing and administering procedures and requirements to prevent the contamination of water supplies by chemigation.100/ The requirements include equipment design, installation, and operation standards, performance monitoring standards, and water protection measures.101/ As part of

---

95/ Iowa Code § 206.14(2)(b).

96/ Iowa Code § 206.14(3).

97/ Iowa Code §§ 206A.1(2), (3), 206A.2(1).

98/ Iowa Code § 206A.2(1).

99/ Iowa Code § 206A.2(1).

100/ Iowa Code § 206A.

101/ Iowa Code § 206A.4(1).

this enforcement responsibility, the DALS is required to conduct certain inspections of chemigation systems that it believes are violating applicable requirements. A person may not refuse access for such an inspection or impede or obstruct the inspection if the DALS presents a search warrant and proper credentials.[102/]

A chemigation permit holder must post a notice to the public warning that chemigation is occurring and identifying any restricted use chemicals applied in the irrigation distribution system.[103/] A permit holder or applicator also must report to the DALS any actual or suspected case of contamination related to the use of chemigation.[104/] The DALS must investigate any such report, and take all action necessary to product the public, including establishing a contamination cleanup plan. The permit holder is required by law to implement the DALS-approved cleanup plan.[105/]

Any applicator applying fertilizers, pesticides, or plant growth regulators by chemigation must be certified by the DALS, and must meet certification renewal and continuing education requirements that vary with the activities performed by the applicator.[106/] The DALS may suspend or revoke the certificate of any applicator applying chemicals by chemigation in violation of chapter 206A or failing to report contamination from chemigation which was known or should have been known to the applicator.[107/]

Civil penalties generally ranging from $100 to $1,000 per offense may be imposed for violation of the chemigation requirements established under

---

102/ Iowa Code § 206A.3.

103/ Iowa Code § 206A.6.

104/ Iowa Code § 206A.7.

105/ Iowa Code § 206A.7.

106/ Iowa Code §§ 206A.5(1), (2).

107/ Iowa Code § 206A.5(3).

chapter 206A.[108] Each day that a violation continues generally will constitute a separate offense. These penalties are in addition to those that may be imposed under chapters 200 and 206 and other provisions of Iowa law.[109] The DALS also has authority to suspend or revoke chemigation permits and certifications, and to seek injunctive relief to enjoin actions relating to chemigation activities that may threaten public safety.[110]

11.4 Agricultural Liming Materials

Agricultural liming materials are regulated under chapter 201 of the Iowa Code. The term "agricultural lime" (or "aglime") is defined as all calcium or magnesium products sold for agricultural purposes in the oxide, hydrate, or carbonate form.[111]

11.4.1 License and Use Requirements

Before any person may sell, offer for sale, or dispose of aglime to be used for soil fertility or correcting soil acidity, an annual license from the DALS must be obtained.[112] The license must be renewed by January 31 of each year. The license application must include the name of the manufacturer or shipper, its principal office, and the name, brand, or trademark under which the aglime will be sold.[113]

Any aglime sold, offered, or exposed for sale in Iowa must be analyzed for its effective calcium carbonate equivalent per ton (ECCE) under the methodology

---

108/ Iowa Code § 206.12(1).

109/ Iowa Code § 206A.12(4).

110/ Iowa Code § 206A.10.

111/ Iowa Code § 201.1(1).

112/ Iowa Code § 201.2.

113/ Iowa Code § 201.2.

established by state law.[114] The law requires a variety of sampling procedures -- by both the producer and the DALS -- from the producer's production site to determine the ECCE of any aglime being produced for distribution.[115] Before a producer may distribute the aglime, the ECCE effectiveness must be certified by the DALS (or the Agricultural Stabilization and Conservation Service (ASCS) in some circumstances) based on the results of samples meeting the requirements of Iowa Code chapter 201[116]. The DALS Secretary is required to provide the necessary certification under a schedule mandated by law.[117]

11.4.2 Labeling Requirements

All agricultural liming materials sold, offered, or exposed for sale must have affixed to their bill of lading or other instrument of sale, shipping, or delivery the ECCE certification of the DALS; the name, brand, or trademark under which the aglime is sold; and the name and principal location of the manufacturer, producer or shipper.[118] Any person who sells, offers for sale, exposes for sale, or distributes any aglime material that does not comply with state requirements, contains an inaccurate ECCE certification, or is adulterated or who makes any false report concerning any aglime violates chapter 201.[119] Such a violation is punishable as a simple misdemeanor and by revocation of the person's aglime license.[120]

114/ Iowa Code § 201.5.

115/ Iowa Code §§ 201.5-201.9.

116/ Iowa Code § 201.10.

117/ Iowa Code §§ 201.8, 201.9.

118/ Iowa Code § 201.10.

119/ Iowa Code § 201.11.

120/ Iowa Code § 201.11.

11.5 Agricultural Chemical Contamination

The Iowa Hazardous Conditions Act is designed to prevent, abate, and control releases of hazardous substances in the environment.121/ The Act, which is discussed in chapter 5 of this handbook, is applicable to "contamination" caused by pesticides or fertilizers.122/ The term "contamination" is defined as the presence of one or more pesticides or the presence of fertilizer in soil or groundwater at levels above those that would result from normal field application rates or above background levels.123/

Iowa Code § 455B.601 directs the Department of Natural Resources (DNR) to create "action levels" -- essentially triggers based on the seriousness of contamination -- for determining necessary responses to agricultural chemical-related contamination. Iowa Code § 455B.601 requires the DNR to classify contamination sites as high, medium, or low priority. The classifications generally are based on the degree and type (e.g., groundwater, soil) of contamination measured in relation to action levels established by the DNR, as well as other pertinent factors such as proximity of the contamination to drinking water, surface water, or ecologically sensitive areas.124/

Whenever agricultural chemical contamination is discovered, the responsible person under Iowa Code chapter 455B must submit for DNR approval an initial site plan addressing the extent and potential migration of contamination, site characteristics, possible drinking water impacts, and a recommended priority classification.125/ If the contamination is classified by the DNR as a high or medium priority site, then further investigation must be conducted to determine the extent

---

121/ Iowa Code §§ 455B.381-455B.399.

122/ Iowa Code § 455B.601.

123/ Iowa Code § 455B.601(1)(a)(2).

124/ Iowa Code § 455B.601(1)(b).

125/ Iowa Code § 455B.601(c).

of remediation which should be undertaken.[126] The law establishes specific corrective action measures that generally must be implemented. These measures vary depending on the type of contamination present and the priority the site is assigned.[127]

126/ Iowa Code § 455B.601(1)(c)(5).

127/ Iowa Code § 455B.601(1)(d).

CHAPTER TWELVE

BIOTECHNOLOGY: GENETICALLY ENGINEERED ORGANISMS

12.0 Introduction

Genetically engineered organisms are primarily regulated under federal law in Iowa. With the growth of biotechnology and its importance to agriculture, federal regulation in Iowa has focused on the release of genetically engineered agricultural plants into the environment through field tests.[1] Nonetheless, regulation with respect to the commercial release of genetically engineered plants as well as other organisms is an increasing area of activity.

12.1 Overview of Federal Biotechnology Regulation

In 1986, the U.S. Environmental Protection Agency (EPA) issued a policy statement declaring that microorganisms and their DNA molecules are chemical substances and, therefore, the EPA had broad powers to regulate genetically engineered organisms under the Toxic Substance Control Act (TSCA).[2] This regulatory power has developed amid controversy, and it is still evolving.

Not all organisms are subject to EPA regulation. Many organisms, such as microbes used as foods, food additives, drugs, cosmetics, medical devices, and pesticides, are not subject to EPA regulation because of statutory exemptions (i.e., coverage by other federal agencies).[3] The EPA generally takes the view that organisms which contain genetic material from organisms from a different genera are subject to EPA regulation under the TSCA.[4] Currently, the EPA requires the creators of genetically engineered organisms to comply with specific TSCA regulations. Chief among these regulations are the premanufacture notice (PMN)

1/ See McGarity, Federal Regulation of Agricultural Biotechnologies, 20 U. Mich. J. L. Ref. 1089 (1987).

2/ 51 Fed. Reg. 23,313, 23,324 (June 26, 1986).

3/ 51 Fed. Reg. 23,313, 23,324 (June 26, 1986).

4/ Id. at 23,326.

requirements for new microorganisms and the reporting requirements under TSCA § 8(e), which require that the manufacturer notify the EPA administrator immediately after discovering that a genetically engineered organism poses a substantial risk of injury to public health or the environment.5/

In addition, EPA maintains that the Federal Insecticide, Fungicide, and Rodenticide Act (FIFRA) section 5 authorizes EPA to regulate pesticides derived from genetic engineering.6/ On September 1, 1994, EPA promulgated regulations to clarify when experimental use permits and notifications are required for genetically engineered microbial pesticides.7/

12.2 TSCA Testing Requirements

One of TSCA'S objectives is to require manufacturers to collect data on the health and environmental effects of their products.8/ The EPA may require that this be done whenever there is an unreasonable risk of injury to the public or the environment or when substantial quantities of the organisms are produced with the potential for substantial environmental or human exposure.9/

On September 1, 1994, EPA proposed regulations under TSCA to screen microorganisms before they are introduced into commerce.10/ The proposed regulations would require a person to submit to EPA a Microbial Commercial Activities Notification (MCAN) before manufacturing a new microorganism for general commercial use. The regulations are expected to be finalized in 1995.

5/ See 15 U.S.C. § 2607(e).

6/ 7 U.S.C. § 136c; see also 40 C.F.R. §§ 158.690 (biochemical pesticides), 158.740 (microbial pesticides).

7/ 59 Fed. Reg. 45,600 (Sept. 1, 1994).

8/ 15 U.S.C. § 2601(b)(1).

9/ 15 U.S.C. § 2604(e).

10/ 59 Fed. Reg. 45,526 (Sept. 1, 1994).

Due to the evolving nature of biotechnology regulations and resultant uncertainty surrounding their applicability, creators of genetically engineered organisms should contact the EPA early in the production stages to ascertain whether testing and EPA approval are required. The EPA reviews submissions under the PMN requirements and may allow the manufacturer to proceed on a limited basis under TSCA § 5(e), pending additional testing and gathering of additional information, if the manufacturer's submission provides sufficient information to evaluate the proposed project. EPA testing determinations may be appealed based upon the record before the agency.[11]

12.3 TSCA Preemption

TSCA generally does not preempt the authority of state or local laws, subject to two primary exceptions.[12] First, if the EPA establishes a testing rule under TSCA § 4, state and local testing requirements for the same chemical are prohibited.[13] Second, if the EPA establishes a rule under TSCA §§ 5 or 6, all state and local regulations for that substance are prohibited except for disposal regulations, those that carry out federal law (e.g., the Clean Air Act), or those that ban the use of the substance (other than its manufacturing uses).[14] Other exceptions may be available.

12.4 TSCA Penalties and Enforcement

Under TSCA, civil penalties up to $20,000 per violation per day may be imposed. Civil penalties for violations of EPA regulations are determined by the gravity of the harm and the actor's culpability based upon knowledge of the TSCA

11/ 15 U.S.C. § 2618(c)(1)(B)(i).

12/ 15 U.S.C. § 2617.

13/ 15 U.S.C. § 2617(a)(2)(A).

14/ 15 U.S.C. § 2617(a)(2)(B).

requirement and the actor's control over the violation.15/ Criminal penalties up to $25,000 per day and/or one year imprisonment may be imposed for knowing or willful violations.16/ TSCA also allows for citizen suits against persons in violation of TSCA and against the EPA to force the performance of a non-discretionary act under the TSCA.17/

12.5 FIFRA Experimental Use Permit and Notification Requirements

In September 1994, EPA promulgated regulations under FIFRA governing experimental use permits (EUP).18/ The regulations establish criteria under which an EUP is presumed not to be required and implement a screening procedure that requires notification to EPA before initiation of small-scale testing of certain microbial pesticides.

In general, the FIFRA regulations create a presumption that no EUP is required when the experimental use is limited to laboratory or greenhouse tests, limited field trials under specified circumstances, and certain other tests to assess a pesticide's potential efficacy, toxicity, or other properties.19/ Despite this presumption, EPA requires notification and EPA approval for any person who plans to conduct small-scale testing of certain microbial pesticides for: (1) small-scale tests that involve an intentional environmental introduction of the microbial pesticide; or (2) small-scale tests performed in a facility without adequate containment and inactivation controls as provided in the regulations.20/

Under FIFRA, EPA has authority to issue stop sale, use, or removal orders or

15/ 15 U.S.C. § 2615(a)(2)(B).

16/ 15 U.S.C. § 2615(b).

17/ 15 U.S.C. § 2619.

18/ 59 Fed. Reg. 45,600 (Sept. 1, 1994) (to be codified at 40 C.F.R. §§ 172.3, 172.43-172.59); see also 40 C.F.R. §§ 158.690 (biochemical pesticides), 158.740 (microbial pesticides).

19/ 40 C.F.R. § 172.3.

20/ 40 C.F.R. § 172.45.

seize pesticides that are in violation of EPA requirements.[21] EPA has authority to assess a civil penalty of not more than $5,000 for each FIFRA offense.[22] Criminal penalties up to $50,000 and/or one year imprisonment apply to knowing violations.[23]

12.6 U.S. Food and Drug Administration Regulation of Biotechnology

The U.S. Food and Drug Administration (FDA) also plays a large role in the regulation of new organisms. In May 1992, the FDA promulgated a policy statement on foods and feeds derived from genetically engineered crop plants.[24] Under this policy, the FDA considers new genetic material added to a plant a "food additive" and, therefore, subject to the added substances section 402(a)(1) of the Federal Food, Drug, and Cosmetic Act (FD&C Act).[25] This section permits the seizure of food, or other regulatory action, if it "bears or contains any poisonous or deleterious substance which may render it injurious to health."[26] Foods may be marketed without prior FDA approval, however, if the additive is in compliance with the requirements of section 402 and it is identical or sufficiently similar to a substance "generally recognized as safe" (GRAS).

In addition, in 1992, Congress considered an amendment to the FD&C Act that would have required makers of genetically engineered or modified plants to obtain FDA approval and attach labels to foods derived from such plants. This concept remains under discussion.

21/ 7 U.S.C. § 136k.

22/ 7 U.S.C. § 136l(a); see also 40 C.F.R. § 172.59.

23/ 7 U.S.C. § 136l(b).

24/ 57 Fed. Reg. 22,984 -23,005 (May 29, 1992).

25/ 21 U.S.C. § 342(a)(1).

26/ 21 U.S.C. § 342(a)(1).

12.7 U.S. Department of Agriculture Regulation of Biotechnology

The U.S. Department of Agriculture (USDA) also regulates certain genetically engineered organisms. The Virus-Serum-Toxin Act (VSTA) allows the USDA to regulate microorganisms used in the treatment of domestic animals.[27] Under VSTA, a person may not prepare or distribute a veterinary virus, serum, or toxin that is intended for use in the treatment of a domestic animal unless it has been prepared under, and in compliance with, the regulations prescribed by the USDA, which may include a licensing requirement.[28] Generally, importation of viruses, serums, and toxins may not be done without a permit from the USDA.[29] The USDA also has broad regulatory powers under the Act of Feb. 2, 1903 to prevent disease or the spread of disease of animals.[30]

Finally, the Federal Plant Pest Act (FPPA) authorizes the USDA to regulate organisms that cause plant diseases or otherwise damage plants.[31] Under the FPPA, creators of genetically engineered organisms that are plant pests are required to obtain a permit from the USDA prior to transportation, importation, or release of such an organism.[32] The USDA, however, recently amended their rules to allow an exemption from the formal permit requirements for six listed crops: corn, cotton, potato, soybean, tobacco, and tomato.[33] Instead of the formal permit, these crops (and any additional plant species that the USDA has determined may be safely introduced) require only that a simple notice be given to the USDA, provided that certain other criteria are met.[34]

27/ 21 U.S.C. §§ 151-59.

28/ 21 U.S.C. §§ 151, 154.

29/ 21 U.S.C. § 152.

30/ 21 U.S.C. §§ 111-136a.

31/ 7 U.S.C. §§ 150aa-150jj.

32/ 7 C.F.R. § 340.3 (permit conditions and application requirements).

33/ Id.; see also 58 Fed. Reg. 17044, 17058-59 (March 31, 1993).

34/ 7 C.F.R. § 340.3(b)(2)-(6); see also 58 Fed. Reg. 17044, 17058-59 (March 31, 1993).

CHAPTER THIRTEEN

ASBESTOS, PCB, and LEAD REGULATION

13.0 Introduction

Asbestos, polychlorinated biphenyls (PCBs), and lead have received special attention from federal and Iowa regulators. Iowa law incorporates workplace exposure requirements for asbestos set forth in federal regulations. Because these asbestos exposure standards have broad application, this chapter focuses on workplace requirements. Given the detailed nature of asbestos, PCB, and lead regulation, obtaining the advice of experienced counsel is particularly important when dealing with these issues.

13.1 Agencies Regulating Asbestos, PCBs, and Lead

In Iowa, exposure to asbestos in the workplace is largely governed by federal regulations adopted by the Occupational Safety and Health Administration (OSHA). These OSHA regulations establish a general industry standard[1] and a construction industry standard.[2] In August 1994, OSHA substantially revised these regulations and added a similar shipyard industry standard to be effective October 11, 1994.[3] Iowa has incorporated the federal OSHA regulations into a state program for regulating exposure to asbestos in the workplace.[4] The Iowa Division of Labor Services of the Department of Employment Services enforces this program.[5]

1/ 29 C.F.R. § 1910.1001(a).

2/ 29 C.F.R. § 1926.1101(a).

3/ 59 Fed. Reg. 40,964 (Aug. 10, 1994). Various parties have filed challenges to these regulations. E.g., Max Eubank Roofing Co. v. U.S. Dep't of Labor, Nos. 94-40793 to 94-40797 (5th Cir. 1994); AFSCME v. Secretary of Labor, No. 94-1575 (D.C. Cir. 1994).

4/ See 29 U.S.C. § 667; Iowa Admin. Code § 347-81.2(2) (adopting in its entirety by reference the federal general industry standard 29 C.F.R. § 1910.1001); Iowa Admin. Code § 347-81.1(2) (adopting in its entirety by reference federal construction industry standard formerly at 29 C.F.R. § 1926.58, now at 29 C.F.R. § 1926.1101).

5/ Iowa Code § 84A.2(2). The Iowa Department of Education oversees a program relating to the training of individuals working on asbestos projects. Iowa Code § 88B.13.

Iowa has supplemented the federal regulations for abatement work involving the removal or encapsulation of asbestos-containing materials (ACM).[6] The Iowa DNR implements this program, which is sometimes referred to as the National Emission Standards for Hazardous Air Pollutants (NESHAPs) program based on the federal Clean Air Act regulations adopted by the EPA relating to asbestos. EPA Region VII continues to monitor this program. Iowa employee right-to-know rules administered by the Labor Commissioner within the Department of Employment Services also apply to employees routinely exposed to asbestos. PCB and lead regulation is largely carried out by the EPA through federal regulation.

13.2 Asbestos Terminology and Background

Asbestos is a naturally occurring mineral that was widely used in building construction until the mid-1970s. "Asbestos" includes chrysotile, amosite, crocidolite, tremolite asbestos, anthophyllite asbestos, actinolite asbestos, and any of these minerals that have been chemically treated or altered.[7] Although the most common short-hand term for asbestos-containing material is ACM, asbestos regulations sometimes refer to regulated asbestos-containing material (RACM), or asbestos-containing building material (ACBM). These terms generally mean material that contains more than one per cent asbestos by weight.[8] Presumed asbestos-containing material (PACM) generally is thermal system insulation and surfacing material found in buildings constructed no later than 1980.[9] Asbestos requirements sometimes are based on the presence of "friable" ACM, which is

---

6/ Iowa Code §§ 88B.1-88B.13; Iowa Admin. Code §§ 347-81.1 to 82.11. Any business entity or individual involved in the removal or encapsulation of asbestos is subject to this Act. Iowa Admin. Code § 347-82.1.

7/ 15 U.S.C. § 2642(3); 29 C.F.R. § 1910.1001(b); 29 C.F.R. § 1915.1001(b); 29 C.F.R. § 1926.1101(b); 40 C.F.R. § 61.141; Iowa Admin. Code § 347-82.1(88B).

8/ 15 U.S.C. § 2642(4); 29 C.F.R. § 1910.1001(b); 40 C.F.R. § 61.141; Iowa Admin. Code § 281-96.3(1).

9/ 29 C.F.R. §§ 1910.1001(b), 1915.1001(b), 1926.1101(b).

defined in the Iowa asbestos abatement regulations as any material containing more than one percent asbestos by weight that hand pressure can crumble, pulverize, or reduce to powder when dry.10/

The OSHA asbestos standards generally govern employers whose employees are or may reasonably be expected to be exposed to concentrations of airborne asbestos fibers exceeding certain minimum levels, known as "Permissible Exposure Limits" (PELs) consisting of a "Time-Weighted Average limit" (TWA) and an "Excursion limit." OSHA has established uniform PELs in the general industry, construction industry, and shipyard industry standards as follows:11/

- **TWA** -- The employer shall ensure that no employee is exposed to an airborne concentration of asbestos in excess of 0.1 fiber per cubic centimeter of air (0.1 f/cc) as an eight-hour time-weighted average as determined by the method prescribed in the regulations.

- **Excursion limit** -- The employer shall ensure that no employee is exposed to an airborne concentration of asbestos in excess of 1.0 fiber per cubic centimeter of air (1 f/cc) as averaged over a sampling period of thirty minutes.

These uniform PELs are more stringent than the standards in effect before October 1994. The term "action level" (typically half of a PEL) was formerly used in the OSHA standards as a lower threshold level triggering various obligations, but OSHA has deleted this term because in many cases, asbestos levels below the more stringent new PELs cannot be reliably measured, and duties tied to an action level might therefore be triggered by measurements of dubious accuracy.12/ Instead of a numerical action level, the revised OSHA regulations impose duties on employers for training and medical surveillance triggered by exposure to ACM or PACM or by the type of work being done. In addition, specified work practices are required regardless of measured exposure levels.

10/ Iowa Admin. Code § 347-82.1; see also 15 U.S.C. § 2642(6); 40 C.F.R. § 61.141.

11/ 29 C.F.R. § 1910.1001(c); 29 C.F.R. § 1915.1001(c); 29 C.F.R. § 1926.1101(c).

12/ 59 Fed. Reg. 40,964, at 40,974.

The general industry standard applies to workplaces in all industries, except construction and shipyard work, and governs the basic requirement that an employer maintain a workplace free of asbestos hazards.13/ The construction industry standard applies to construction employers engaged in a wide variety of construction and abatement activities involving asbestos.14/ The shipyard industry standard applies to all shipyard employment work.15/ The basis for the application of separate standards is the recognition of the differences in asbestos exposure and workplace conditions in general industry, construction, and shipyard worksites.16/

The construction and shipyard standards employ the same basic framework of the general industry standard. OSHA has eliminated deviations from this framework in the construction industry standard for "short-term, small-scale operations" in favor of four classifications of construction activity that are matched with increasingly stringent control requirements.17/ These "classes" of asbestos work are discussed below in connection with the construction industry standard.

13.3 General Industry Standard

The general industry standard applies to all occupational exposures to asbestos, except construction and shipyard work. The standard includes employee exposures that result from the employee's presence in a building where asbestos products have been installed, regardless of an employee's involvement with asbestos products.18/ Although building owners formerly were liable only as

---

13/ 29 C.F.R. § 1910.1001(a).

14/ 29 C.F.R. § 1926.1101(a).

15/ 29 C.F.R. § 1915.1001(a).

16/ See 59 Fed. Reg. 40,964, at 40,968, 41,018 (Aug. 10, 1994); 51 Fed. Reg. 22,612-22,613 (1986).

17/ 59 Fed. Reg. 40,964, at 40,969 (Aug. 10, 1994).

18/ See 51 Fed. Reg. 22,612, at 22,678 (1986); see also 59 Fed. Reg. 40,964, at 40,972-40,973 (Aug. 10, 1994) (indicating that revised general industry standard focuses on four industry segments: brake and clutch repair, custodial workers, primary asbestos manufacturers, and secondary

employers, not as owners of buildings where asbestos is present,[19/] certain obligations in the revised regulations apply to building/facility owners. A building/facility owner is defined as the legal entity, including a lessee, which exercises control over management and recordkeeping functions relating to a building and/or facility in which activities covered by the general industry standard take place.[20/] Employers subject to the general industry standard are defined broadly to include any person employing one or more employees.[21/]

13.3.1 Monitoring Exposure to Asbestos

Every employer with a workplace or work operation covered by the general industry standard may be obligated to perform initial monitoring of employees who are or may reasonably be expected to be exposed to airborne concentrations of asbestos at or above the PELs.[22/] The regulations, however, do not indicate when it may "reasonably be expected" that employee exposure would exceed a PEL. Nonetheless, office buildings generally have asbestos concentrations well below the PEL. Thus, initial monitoring would rarely be required unless conditions exist that may expose employees to asbestos above a PEL.[23/] For example, visible deterioration of ACM and construction or maintenance activities that would substantially disturb

---

asbestos manufacturers).

19/ 51 Fed. Reg. 22,612, at 22,678 (1986).

20/ 29 C.F.R. §§ 1910.1001(b); see also 59 Fed. Reg. 40,964, at 41,013 (Aug. 10, 1994).

21/ 29 C.F.R. §§ 1910.2(c), 1926.32(k). This definition does not include any State or political subdivision of a State. Asbestos abatement regulations applicable to these governmental entities are at 40 C.F.R. §§ 763.120-763.126.

22/ 29 C.F.R. § 1910.1001(d)(2)(i).

23/ See 51 Fed. Reg. 22,612, at 22,683 (1986).

ACM may trigger monitoring requirements.[24/] Initial monitoring is not required if the employer can demonstrate that asbestos will not be released above PELs.[25/]

The employer must also conduct periodic monitoring of asbestos concentrations at least every six months if the initial monitoring shows that a PEL is exceeded or if it is reasonably foreseeable that employee exposure will exceed a PEL.[26/] When the initial or periodic monitoring shows airborne asbestos concentrations below PELs, monitoring may be discontinued unless there is a change in the workplace that may result in new or additional exposure.[27/] OSHA rules set out the procedures for monitoring.[28/] The employer must notify the affected employees within specified time frames of the results of any monitoring performed under the general industry standard in writing or by posting of results.[29/] The employer must provide affected employees or their designated representatives an opportunity to observe any monitoring of employee exposure to asbestos.[30/]

13.3.2 Medical Surveillance

If monitoring indicates that an employee has been or will be exposed to asbestos concentrations above a PEL, the employer may be required to institute a medical surveillance program for all employees subject to such exposure.[31/] Under the medical surveillance program, the employee or former employee is to be

---

24/ 51 Fed. Reg. 22,612, at 22,683-684 (1986).

25/ 29 C.F.R. § 1910.1001(d)(2)(ii)-(iii).

26/ 29 C.F.R. § 1910.1001(d)(3).

27/ 29 C.F.R. § 1910.1001(d)(4),(5).

28/ 29 C.F.R. § 1910.1001(d)(1),(6); 29 C.F.R. § 1910.1001 App. A.

29/ 29 C.F.R. § 1910.1001(d)(7).

30/ 29 C.F.R. § 1910.1001(n).

31/ 29 C.F.R. § 1910.1001(l)(1).

examined annually by a licensed physician before, during, and after exposure.32/ The physician conducting the examination must issue a written opinion on the health risks the employee runs due to exposure33/ and must inform the employee of the examination results.34/

13.3.3 Regulated Asbestos Areas

An employer must establish regulated areas where asbestos concentrations exceed or can reasonably be foreseen to exceed a PEL.35/ Regulated areas are to be separated, and employees entering these areas must use a respirator and wear protective clothing.36/ Employers are required to post warning signs saying: "DANGER/ ASBESTOS/ CANCER AND LUNG DISEASE HAZARD/ AUTHORIZED PERSONNEL ONLY/ RESPIRATORS AND PROTECTIVE CLOTHING ARE REQUIRED IN THIS AREA.".37/

Employers are also required to provide clean change rooms, showers, and lunch rooms.38/ Employers must ensure that employees shower and do not leave the workplace wearing protective clothing.39/ Employers must ensure that employees do not smoke, eat, drink, or chew gum in regulated areas.40/ Other requirements apply to lunch room facilities.41/

32/ 29 C.F.R. § 1910.1001(l)(1)-(4).

33/ 29 C.F.R. § 1910.1001(l)(7)(i)(A)-(B).

34/ 29 C.F.R. § 1910.1001(l)(7)(i)(C)-(D).

35/ 29 C.F.R. § 1910.1001(e)(1).

36/ 29 C.F.R. § 1910.1001(e)(2)-(4),(h)(1).

37/ 29 C.F.R. § 1910.1001(j)(3).

38/ 29 C.F.R. § 1910.1001(i).

39/ 29 C.F.R. § 1910.1001(i)(2)(iii).

40/ 29 C.F.R. § 1910.1001(e)(5), (i)(4).

41/ 29 C.F.R. § 1910.1001(i)(3).

13.3.4 Employee Information and Training

As stated above, employees must be notified of monitoring results and, if monitoring results indicate that PELs have been exceeded, the corrective action being taken. The 1994 revisions to the OSHA regulations substantially expanded the obligations of employers and building owners regarding communication about hazards to employees. These requirements include obligations to exercise "due diligence" to inform employers and employees about the presence and location of ACM and PACM.

The new provisions obligate building and facility owners to maintain records of all information required to be provided regarding communication of hazards to employees and information otherwise known to the owner concerning the presence, location, and quantity of ACM and PACN in the building/facility.42/ Records must be kept for the duration of ownership and transferred to successive owners. Building and facility owners must inform employers (and employers must inform employees who will perform housekeeping activities in areas that contain ACM or PACM) of the presence and location of ACM and PACM.43/

As part of the hazard communication requirements, warning labels must be affixed to all raw materials, mixtures, scrap, waste, debris, and other products containing asbestos fibers, or to their containers.44/ Employers must institute a training program for employees exposed to asbestos concentrations above a PEL.45/ Training is to be conducted before the employee's initial assignment and at least annually thereafter. The training must inform the employee of:

1. The health affects associated with exposure to asbestos;
2. The relationship between smoking and exposure to asbestos producing lung cancer;

42/ 29 C.F.R. § 1910.1001(j)(2); see also 59 Fed. Reg. 40,964, at 41,013-41,014 (Aug. 10, 1994).

43/ 29 C.F.R. § 1910.1001(j)(2)(iii).

44/ 29 C.F.R. § 1910.1001(j)(4), (j)(6).

45/ 29 C.F.R. § 1910.1001(j)(7).

3. The circumstances under which exposure could result;
4. The procedures implemented to protect employees subject to exposure; and
5. Other information relating to protective equipment and the medical surveillance program.46/

The employee must also be given access to OSHA standards, training materials, and smoking cessation program materials.47/ Similar training requirements apply for employees who perform housekeeping operations in a facility which contains ACM or PACM.48/

13.3.5 Compliance with Permissible Exposure Limits (PELs)

Employers are required to institute engineering controls and work practices to reduce and maintain employee exposure to airborne asbestos below the PELs when feasible.49/ Work practices, such as local ventilation systems, are required for tools, like drills, that would release asbestos fibers above the PELs.50/ Asbestos should be wet when worked with to avoid asbestos emissions into the air above the PELs.51/

When engineering and work practices cannot bring asbestos exposure below the PELs, respirators meeting OSHA standards must be supplied and used.52/ Where the PEL is exceeded, employers must establish and implement a written compliance program to reduce employee exposure by engineering and work practice controls and by the use of mandatory respiratory protection.53/

46/ 29 C.F.R. § 1910.1001(j)(7)(iii).

47/ 29 C.F.R. § 1910.1001(j)(7)(v).

48/ 29 C.F.R. § 1910.1001(j)(7)(iv).

49/ 29 C.F.R. § 1910.1001(f)(1).

50/ 29 C.F.R. § 1910.1001(f)(1)(v).

51/ 29 C.F.R. § 1910.1001(f)(1)(vi),(viii).

52/ 29 C.F.R. § 1910.1001(f)(1)(ii)-(iii),(g).

53/ 29 C.F.R. § 1910.1001(f)(2).

13.3.6 Protective Clothing and Equipment

When employees are exposed to asbestos concentrations in excess of a PEL, employers must supply and ensure that employees use appropriate protective work clothing and equipment.54/ This protective clothing and equipment must be removed only in designated change rooms and stored in closed containers to prevent the dispersion of asbestos into the air.55/ The clothing and equipment must be cleaned weekly and replaced as appropriate to maintain its effectiveness. Laundering is to be done so as to prevent the release of asbestos above the PELs.56/

13.3.7 Asbestos Housekeeping Requirements

Under the general industry standard, surfaces must be kept free of accumulations of dust and waste containing asbestos, and releases of asbestos products must be promptly cleaned up.57/ Where feasible, only vacuuming (with high-efficiency particulate air (HEPA) filtered equipment) or wet methods are permissible for cleaning.58/ Disposal of waste contaminated with asbestos must be done in sealed, impermeable containers.59/ Specific procedures are set forth for care of asbestos-containing flooring material.60/

13.3.8 Asbestos Recordkeeping Requirements

The general industry standard imposes extensive recordkeeping requirements

---

54/ 29 C.F.R. § 1910.1001(h)(1).

55/ 29 C.F.R. § 1910.1001(h)(2).

56/ 29 C.F.R. § 1910.1001(h)(3).

57/ 29 C.F.R. § 1910.1001(k)(1)-(2).

58/ 29 C.F.R. § 1910.1001(k)(3)-(5).

59/ 29 C.F.R. § 1910.1001(k)(6).

60/ 29 C.F.R. § 1910.1001(k)(7).

on employers.61/ Records of air monitoring measurements of asbestos are to be kept for at least 30 years.62/ The employer must also retain records about medical surveillance programs and employee training.63/ Required records must be made available to OSHA officials and affected employees and former employees for examination and copying.64/

13.4 Construction Industry Standard

In contrast to the general industry standard, the construction industry standard applies to exposures to asbestos in work involving construction, alteration, and/or repair, including painting and decorating.65/ The standard specifically regulates the following:

1. Construction operations involving materials or structures containing asbestos, including demolition or salvage; removal or encapsulation;
2. Alteration, repair, maintenance, or renovation;
3. Installation;
4. Cleanup and housekeeping; and
5. Transportation, disposal, storage, and containment.66/

Thus, the construction industry standard actually applies very broadly.

The construction and general industry standards apply essentially the same basic framework for asbestos regulation. For example, each standard applies the same PELs. The construction industry standard, however, adopts a tiering approach

61/ 29 C.F.R. § 1910.1001(m).

62/ 29 C.F.R. § 1910.1001(m)(1)(iii).

63/ 29 C.F.R. § 1910.1001(m)(3),(4).

64/ 29 C.F.R. § 1910.1001(m)(5).

65/ 29 C.F.R. § 1926.1101(a).

66/ 29 C.F.R. § 1926.1101(a).

under which the stringency of the standard's requirements varies according to the type of construction operation. These classes are defined as follows:[67]

- Class I asbestos work -- activities involving the removal of Thermal System Insulation (TSI) and surfacing ACM and PACM.

- Class II asbestos work -- activities involving the removal of ACM that is not TSI or surfacing material. This includes, but is not limited to, the removal of asbestos-containing wallboard, floor tile and sheeting, roofing and siding shingles, and construction mastics.

- Class III asbestos work -- repair and maintenance operations where ACM, including TSI and surfacing material, is likely to be disturbed.

- Class IV asbestos work -- maintenance and custodial activities during which employees contact ACM and PACM and activities to clean up waste and debris containing ACM and PACM.

TSI means ACM (more than one percent asbestos) applied to pipes, fittings, boilers, breeching, tanks, ducts, or other structural components to prevent heat loss or gain.

13.4.1 Monitoring Exposure to Asbestos

The construction industry standard, like the general industry standard, requires employers to conduct initial monitoring of employee exposure to asbestos.[68] For Class I and Class II operations, construction employers are required to conduct daily periodic monitoring unless the employer has made a negative exposure assessment under specified procedures.[69] For operations other than Class I and Class II asbestos work, periodic monitoring is required where exposures are expected to exceed a PEL. However, monitoring may be discontinued under

---

67/ 29 C.F.R. § 1926.1101(b).

68/ 29 C.F.R. § 1926.1101(f)(2)(i).

69/ 29 C.F.R. § 1926.1101(f)(3).

specified circumstances.70/ As with general industry standards, employees are to be notified of monitoring results and allowed to observe the monitoring.71/

13.4.2 Medical Surveillance

The construction industry standard mandates a medical surveillance program similar to the general industry standard. Construction employers are required to institute a medical surveillance program for affected employees: (1) if asbestos monitoring indicates that an employee will be exposed to asbestos concentrations above the PELs or are engaged in Class I, II, or III work for 30 or more days per year; or (2) if an employee is required to wear a negative-pressure respirator.72/

Medical examinations of employees covered by the surveillance program must be performed at the following times: (1) prior to assignment to an area where respirators must be worn, (2) within ten working days following the thirtieth day of exposure when the employee has been exposed to asbestos above a PEL for 30 or more days per year, (3) at least annually thereafter, and (4) at such other times the examining physician determines necessary.73/

13.4.3 Regulated Asbestos Areas, Recordkeeping, and Housekeeping

As with general industry employers, construction employers must establish regulated areas where airborne asbestos concentrations exceed or can reasonably be expected to exceed a PEL and also where Class I, II, or III asbestos work will be conducted.74/ Similar requirements exist on separating and allowing access to regulated areas, providing protective equipment and clean areas, prohibiting

---

70/ 29 C.F.R. § 1926.1101 (f)(4).

71/ 29 C.F.R. § 1926.1101(f)(5), (n)(7).

72/ 29 C.F.R. § 1926.1101(m)(1)(i).

73/ 29 C.F.R. § 1926.1101(m)(2).

74/ 29 C.F.R. § 1926.1101(e)(1).

smoking and other activities within the regulated areas, and posting warning signs around these areas.[75] An employer must ensure that asbestos work performed within regulated areas is supervised by a competent person.[76]

Recordkeeping and housekeeping requirements under the construction industry standard are essentially identical to those under the general industry standard.[77] For a building owner that has communicated information concerning the identification, location, and quantity of ACM and PACM, written records of the notifications and their content must be maintained by the building owner for the duration of ownership and transferred to successive owners of the building.[78]

13.4.4 Multi-Employer Worksites

On multi-employer worksites, an employer performing work requiring the establishment of a regulated area must inform other employers on the site of the nature of the work with ACM or PACM, the requirements governing regulated areas, and the measures taken to protect employees from asbestos exposure.[79] Additional obligations for multi-employer sites, including general contractor responsibilities, are set forth in the regulations.

13.4.5 Employee Information and Training -- Hazard Communication

Construction employers must institute a training program for all employees who install asbestos-containing products and for all employees who perform Class I-IV asbestos operations, even though this work may not necessarily result in

---

75/ 29 C.F.R. § 1926.1101(e)(2)-(5).

76/ 29 C.F.R. § 1926.1101(e)(6), (o).

77/ 29 C.F.R. § 1926.1101(l),(n).

78/ 29 C.F.R. § 1926.1101(n)(6).

79/ 29 C.F.R. § 1926.1101(d).

exposure to asbestos above the PEL.80/ The required content of the program is set forth in the regulations, generally depending on the class of work performed.81/ In addition, the competent person charged with overseeing abatement operations must be specially trained in the handling of asbestos.82/

In the regulations adopted in 1994, hazard communication requirements are established for building and facility owners. These hazard communication requirements include obligations to identify ACM and PACM and to notify tenants who will occupy areas containing ACM or PACM and certain employers and employees.83/ At the entrance to mechanical rooms that contain TSI and surface ACM or PACM, the building owner must post signs that identify the presence, location, and appropriate work practices to prevent disturbing the ACM or PACM.84/ Warning signs must demarcate regulated areas and warning labels containing information as required under the general industry standard must be placed on all products containing asbestos, except where exposure will be less than the PELs or asbestos represents less than 0.1 percent of the product by weight.85/

13.4.6 Methods of Compliance

In contrast to standards based on compliance with PELs, construction employers must institute engineering controls and work practices to reduce and maintain employee exposure to asbestos based on the type of work being performed regardless of the levels of exposure.86/ These controls and practices include local

---

80/ 29 C.F.R. § 1926.1101(k)(8).

81/ See 29 C.F.R. § 1926.1101(k)(8).

82/ 29 C.F.R. § 1926.1101(o)(4).

83/ 29 C.F.R. § 1926.1101(k)(1)(ii)(D).

84/ 29 C.F.R. § 1926.1101(k)(5).

85/ 29 C.F.R. § 1926.1101(k)(6), (7).

86/ 29 C.F.R. § 1926.1101(g).

and general ventilation systems, vacuuming, isolation of asbestos operations, wet methods, and leakproof containers.87/ Some work practices are flatly prohibited.

The regulations set forth detailed work practices based on the class of work being performed, including negative pressure enclosures (NPE) for various Class I jobs and stringent requirements based on jobs involving the removal of more than 25 linear or 10 square feet of TSI or surfacing material..88/ Additional regulations governing hygiene facilities and practices, such as decontamination areas, shower areas, clean change areas, and lunch rooms, also are based on the class of work being performed.89/

Respirators meeting OSHA specifications and other protective equipment must be supplied to and used by employees when engineering controls and work practices do not bring asbestos exposure below the PELs and in specified Class I-IV jobs.90/ The engineering controls, work practices, and respirators employed must be set forth in a written program.91/

13.5 Shipyard Industry Standard

In the regulations adopted in August 1994, OSHA added extensive provisions governing the shipyard industry which are similar to the framework of the general industry standard and construction industry standard.92/ Like the construction industry standard, these regulations are based on the nature of the work operation involving asbestos exposure rather than the employers overall business. Because

---

87/ 29 C.F.R. § 1926.1101(g)(1)(i)-(v).

88/ 29 C.F.R. § 1926.1101(g).

89/ 29 C.F.R. § 1926.1101(j).

90/ 29 C.F.R. § 1926.1101(g)(2)(v), (h).

91/ 29 C.F.R. § 1926.1101(h)(3) (incorporating written respirator protection requirements in 29 C.F.R. § 1910.134(b)).

92/ 59 Fed. Reg. 40,964 (Aug. 10, 1994) (to be codified at 29 C.F.R. § 1926.1001).

the shipyard industry is not extensive in Iowa, these regulations are not discussed in detail here.

13.6 Asbestos Penalties Under OSHA

An employer that violates the requirements of the general industry, shipyard industry, or the construction industry standard is subject to civil and criminal penalties.[93/] The maximum penalty for each serious or nonserious violation is $7,000.[94/] An employer who willfully or repeatedly violates the requirements of the standards, however, may be subject to a maximum penalty of $70,000 for each violation.[95/] The minimum fine for a willful violation is $5,000.

In addition, if a willful violation of the standards causes death to any employee, the employer will upon conviction be punished by a maximum fine of $10,000 or imprisonment not to exceed six months, or both.[96/] The same penalties apply to an employer that knowingly makes any false statements, representations, or certifications in any records or other documents required under the standards.[97/] An employer that fails to correct a violation for which a citation has been issued within the time permitted for its correction may be assessed a civil penalty of not more than $7,000 for each day during which such failure or violation continues.[98/] An employer is also subject to a civil penalty of up to $7,000 for each violation of any of the posting requirements required by the standards.[99/] Asbestos abatement projects by state and local governments, which are subject to regulation under 40

93/ 29 U.S.C. § 666; see also 29 U.S.C. § 655; 29 C.F.R. pts. 1910, 1926.

94/ 29 U.S.C. § 666(b)-(c).

95/ 29 U.S.C. § 666(a).

96/ 29 U.S.C. § 666(e). For any subsequent violation, the penalty increases to a fine of not more than $20,000 or imprisonment for not more than one year, or both.

97/ 29 U.S.C. § 666(g).

98/ 29 U.S.C. § 666(d).

99/ 29 U.S.C. § 666[(h)](i).

C.F.R. § 763.120 rather than OSHA, are backed by penalties under the Toxic Substances Control Act (TSCA)[100]/

As noted above, the Iowa Division of Labor Services has incorporated by reference the OSHA general industry standard and construction industry standard into Iowa rules promulgated pursuant Iowa Code chapter 88B. Thus, violations also may be subject to a civil penalty to the State of Iowa of not more than $5,000 for each violation and criminal penalties for subsequent willfull violations.[101]/

13.7 Asbestos Abatement, Removal, And Encapsulation In Iowa

13.7.1 Covered Activities

In addition to the specific construction industry standards governing worker exposure levels during asbestos abatement projects, more general NESHAP regulations at 40 C.F.R. § 61.145 apply to demolition and renovation projects involving regulated asbestos-containing material or RACM.[102]/ These NESHAP regulations impose notification requirements and set forth procedures for asbestos emission control. In most circumstances, stricter requirements apply to abatement projects involving the following:

1. At least 80 linear meters (260 linear feet) on pipes or at least 15 square meters (160 square feet) on other facility components; or

2. At least one cubic meter (35 cubic feet) off facility components where the length or area could not be measured previously.[103]/

Iowa regulates asbestos abatement activities under Iowa Code chapter 88B, referred to as the Asbestos Removal and Encapsulation (the Act), and implementing

---

100/ 15 U.S.C. § 2614; 40 C.F.R. § 763.125; see also 59 Fed. Reg. 54,746 (Nov. 1, 1994) (proposing amendments to regulations).

101/ Iowa Code § 88B.12; Iowa Admin. Code § 347-81.2(2) (adopting general industry standard); Iowa Admin. Code § 347-81.1(2) (adopting construction industry standard).

102/ Iowa has adopted this section by reference. Iowa Admin. Code § 567-23.1(3).

103/ 40 C.F.R. § 61.145(a).

regulations promulgated by the Iowa Division of Labor Services of the Department of Employment Services.[104] Asbestos projects subject to the Act's provisions include the removal or encapsulation of friable asbestos material or other releases of asbestos such as by the operation of hand-operated or power-operated tools which would produce or release fibers of asbestos or other substantial alteration of asbestos containing nonfriable material.[105] The provisions impose licensing, certification, and reporting requirements and also set forth standards to be followed when conducting abatement activities. Contrary to popular conceptions about asbestos abatement, there is no general obligation to remove asbestos from buildings under federal or Iowa law.[106]

13.7.2 Iowa Asbestos Abatement Permit for Business Entities

A business entity in Iowa that engages in the removal or encapsulation of asbestos is required to hold a permit for that purpose.[107] Also, an Iowa state agency or political subdivision will not accept a bid in connection with any asbestos project from a business entity that does not hold a permit from the Iowa Division of Labor Services at the time the bid is submitted.[108] Employers performing in-house abatement work are exempt from the permit requirement, except that these employers are required to provide training programs for their employees.[109] To

---

104/ See Iowa Code §§ 88B.1-88B.13; Iowa Admin. Code rr. 347-81.1-82.11.

105/ Iowa Code § 88B.1; Iowa Admin. Code § 347-82.1 (defining "asbestos project").

106/ EPA, Asbestos NESHAP Clarification of Intent, 58 Fed. Reg. 51,784 (Oct. 5, 1993); EPA, Managing Asbestos in Place: A Building Owner's Guide to Operations and Maintenance for Asbestos-Containing Material (July 1990).

107/ Iowa Code § 88B.2.

108/ Iowa Code § 88B.11.

109/ Iowa Code § 88B.2. This training is subject to review and approval upon inspection by the Division. In addition, a local education agency that assigns an employee to perform small-scale, short-duration operation maintenance and repair activities is also exempt from the permit requirement. Iowa Admin. Code § 82.6(6).

obtain a permit, a business entity must complete an application to the Division and pay the applicable fee.110/ Permits expire on the first anniversary of their effective date, unless renewed for another one-year term.111/ The Labor Commissioner may waive the permit requirement in an emergency situation.112/

13.7.3 Licensing of Asbestos Workers

Each person working on an asbestos project is required to hold a license issued by the Division.113/ The regulations require at least one licensed supervisor or contractor at an asbestos project at all times while work is in progress.114/ Separate license and application procedures exist for workers, supervisors and contractors, inspectors, management planners, and abatement project designers.115/ The requirements for the license include successful completion of training courses as specified by the Iowa Department of Education.116/

110/ Iowa Code § 88B.4; Iowa Admin. Code § 347-82.3.

111/ Iowa Code § 88B.6.

112/ Iowa Code § 88B.9(1). The Commissioner may also waive the requirement for a permit for a business entity not primarily engaged in the removal or encapsulation of asbestos if worker protection requirements are met or an alternative procedure is approved. Iowa Code § 88B.9(3).

113/ Iowa Code § 88B.10.

114/ Iowa Admin. Code § 347-81.4. The "supervisor/contractor" classification includes those persons who provide supervision to workers engaged in the asbestos removal, encapsulation, enclosure, and repair. Asbestos workers must have access to at least one supervisor/contractor throughout the duration of an asbestos project. Iowa Admin. Code § 347-81.4.

115/ Iowa Admin. Code § 347-82.6.

116/ Iowa Admin. Code § 347-82.6. All persons seeking a license must complete an initial three- or four-day training course and an annual refresher course. See Iowa Admin. Code §§ 281-96.10 to 96.12. In addition, workers, supervisors, and contractors are required to provide a certificate from a physician indicating the applicant is physically capable of working while wearing a respirator and a statement that a respirator fit test was successfully passed. Iowa Admin. Code § 347-82.6(1)-(2); see also American Asbestos Training Center, Ltd. v. Eastern Iowa Comm. College, 463 N.W.2d 56 (Iowa 1990).

13.7.4 Reporting Asbestos Abatement Work

Business entities performing abatement work must notify the Division as well as the DNR and EPA Region VII at least ten days before beginning an asbestos project.117/ The notification must include information regarding the business entity, the location of the asbestos project, a description of the structure and asbestos work to be performed, the anticipated dates of the project, and a designation of the disposal site for asbestos material and asbestos contaminated materials.118/

13.7.5 Asbestos Abatement Recordkeeping Requirements

Records must be kept for each asbestos project, including a record of individuals working on the project, the location of and a description of the project, including the amount of asbestos material removed, a summary of the procedures that were used to comply with all applicable standards, and the location of each asbestos disposal site.119/ Business entities must keep these records for at least six years.120/

13.7.6 Asbestos Abatement Plan; Encapsulation Procedures

Business entities must prepare a written asbestos abatement plan describing the equipment and procedures to be used throughout the abatement project.121/

117/ Iowa Admin. Code § 347-82.4. When there is an immediate danger to life, health, or property, the Division must be notified within five days of the initiation of the project. Iowa Admin. Code § 347-82.4(1). The form for notification of DNR and EPA is reprinted following 40 C.F.R. § 61.145.

118/ Iowa Admin. Code § 347-82.4(3).

119/ Iowa Code § 88B.7; Iowa Admin. Code § 347-82.5. The Division also requires a record of the starting and completion dates of each instance of removal or encapsulation and a receipt from the asbestos disposal site.

120/ Iowa Admin. Code § 347-82.5.

121/ Iowa Admin. Code § 347-82.3. As part of the permit application, a business entity is required to provide information that includes a description of the protective clothing and respirators that the business entity will use, a copy of the business entity's respiratory protection program, a description of the site decontamination procedures that the business entity will use, a

Iowa rules also set forth specific work standards applicable to the encapsulation of asbestos.[122] For example, business entities may use encapsulation only when the asbestos, tremolite, anthophyllite or other actinolite fibers are firmly bonded to the underlying surface or the asbestos, tremolite, anthophyllite or actinolite is not accessible for removal. Encapsulation may not be used on material that is deteriorated or delaminated, that shows extensive damage, in areas where contact damage may occur, or areas with actual or probable water damage.[123] Also, sealants must be applied with airless spray equipment at low-pressure settings.[124] All encapsulated asbestos, tremolite, anthophyllite or actinolite sources must be inspected every six months.[125] Previously encapsulated asbestos, tremolite, anthophyllite or actinolite materials must be removed prior to demolition of any building.[126]

13.7.7 Asbestos Abatement Enforcement and Penalties

The Division administers and enforces asbestos removal and encapsulation provisions.[127] At least once a year during an actual asbestos project, the Division is

---

description of the procedures that the business entity will use for handling waste containing asbestos, and a description of the procedures that the business entity will use in cleaning up after completion of the project.

122/ Iowa Admin. Code § 347-81.3. Other encapsulation methods may be used provided the method constructs an impermeable enclosure to prevent friable asbestos, tremolite, anthophyllite or actinolite from becoming airborne. Iowa Admin. Code § 347-81.3(9).

123/ Iowa Admin. Code § 347-81.3(2).

124/ Iowa Admin. Code § 347-81.3(3). Accurate records on the type of solvent used and the nature of the material and substrate encapsulated must be maintained and provided to the owner of the building. Iowa Admin. Code § 81.3(6). Whenever solvent-based sealants are used, the employer must provide an appropriate respirator. Iowa Admin. Code § 81.3(7). Only latex paint with a vehicle content of at least 60% by weight and vehicle resin solids of at least 25% by weight may be used. Iowa Admin. Code § 81.3(8).

125/ Iowa Admin. Code § 81.3(4). Sealants must be reapplied periodically wherever there is a danger of airborne asbestos, tremolite, anthophyllite or actinolite fibers.

126/ Iowa Admin. Code § 81.3(5).

127/ Iowa Code § 88B.3.

required to conduct an on-site inspection examining each business entity's procedures for removing and encapsulating asbestos.[128/] In addition, a business entity is required to make its records available to the Division or the Department of Education at any reasonable time.[129/] The Division also has the authority to reprimand a permittee or licensee and suspend or revoke a permit or license.[130/]

A person or business entity that wilfully violates a provision of the Act or a rule adopted pursuant to the Act is subject to a civil penalty of not more than $5,000 for each violation.[131/] A person or business entity that previously has been assessed a civil penalty and that wilfully violates a provision of the Act or a rule adopted pursuant to the Act, for a first offense, is guilty of a simple misdemeanor and will receive a fine not to exceed $20,000.[132/] For a second or subsequent offense, the business entity or person is guilty of an aggravated misdemeanor and may receive a fine not to exceed $25,000 or a term of imprisonment not to exceed two years, or both.[133/]

NESHAP requirements are enforced by the DNR pursuant to Iowa Admin. Code § 567-23 and by the EPA pursuant to federal Clean Air Act provisions at 42 U.S.C. § 7413(b). Based on this statutory authority, the DNR and the EPA may seek civil penalties of up to $25,000 per violation per day and stricter criminal penalties.

128/ Iowa Code § 88B.3.

129/ Iowa Code § 88B.8.

130/ Iowa Code § 88B.8.

131/ Iowa Code § 88B.12.

132/ Iowa Code § 88B.12(1)(a).

133/ Iowa Code § 88B.12(2)(b).

13.8 Iowa Employee Right-to-Know Rules

In addition to training requirements, Iowa employee right-to-know rules also apply to employees routinely exposed to asbestos.134/ An employer is required to inform employees about the hazardous chemicals to which the employee may be exposed in the workplace, the potential health hazards of the hazardous chemicals, and the proper handling techniques for the hazardous chemicals.135/ Employers are required to develop, implement, and maintain a written hazard communications program for their workplaces pertaining to labels and other forms of warning, material safety data sheets, and employee information and training.136/

13.9 Miscellaneous Asbestos Regulations

In addition to workplace and asbesots abatement regulations, other federal and Iowa environmental statutes also regulate asbestos. For example, under Iowa law, asbestosis is a reportable disease.137/ Like the NESHAPs applicable to asbestos emissions from abatement projects, emissions of asbestos into the air from other sources are governed by air pollution regulations or NESHAPs under the federal Clean Air Act.138/ Effluent limits govern discharges of asbestos into waters.139/

134/ Iowa Code § 89B.1; Iowa Admin. Code 347-110.3(3); Iowa Admin. Code rr. 347-120.1 to 120.11; see also chapter 8 of this handbook.

135/ Iowa Code § 89B.8(1); Iowa Admin. Code § 347-120.6. These rules apply to any chemical which is known to be present in the workplace so that employees may be exposed under normal conditions of use or in a foreseeable emergency. Iowa Admin. Code § 347-110.1(2).

136/ Iowa Admin. Code § 347-120.3. These records are subject to inspection by compliance safety and health officers. Iowa Admin. Code § 347-120.9.

137/ Iowa Admin. Code § 641-1.2. Asbestosis is a specific noninfectious disease required to be reported by physicians and other health practitioners to the Iowa Department of Public Health.

138/ See 42 U.S.C. §§ 7401-7671q; 40 C.F.R. §§ 61.140-61.156 (NESHAPs for asbestos); Iowa Admin. Code § 567-23.1(3) (adopting NESHAPs by reference).

139/ See 33 U.S.C §§ 1251-1387; 40 C.F.R. §§ 427.10-427.116; Iowa Admin. Code § 567-41.3.

The EPA has adopted regulations relating to the reporting of commercial and industrial imports, processing, and use of asbestos under the Toxic Substances Control Act.140/ In 1991, a federal circuit court vacated most of these regulations.141/ In June 1994, EPA revised these regulations, which ban the distribution in commerce of certain specified asbestos-containing products on a schedule phased in over 1990-1997.142/

Congress amended TSCA in 1986 by adopting the Asbestos Hazard Emergency Response Act (AHERA). AHERA requires EPA to establish a program for inspection, management, planning, operations and maintenance activities, and appropriate abatement responses for ACM in schools.143/ In addition, Iowa rules control the abatement of ACM in schools.144/ U.S. Department of Transportation (DOT) regulations governing asbestos-containing waste material establish requirements relating to waste containment and shipping papers.145/ Finally, the EPA has listed asbestos as a hazardous substance for purposes of the Superfund Act discussed in chapter 5 of this handbook.146/

13.10 PCB Regulation

PCBs are compounds that commonly were used in dielectric fluids in transformers and capacitors, in hydraulic fluids, and in solvents because PCBs are

---

140/ See 15 U.S.C. § 2601; 40 C.F.R. §§ 763.60-763.78, 763.160-763.179.

141/ Corrosion Proof Fittings v. EPA, 947 F.2d 1201 (5th Cir. 1991).

142/ 59 Fed. Reg. 33,208 (June 28, 1994).

143/ 15 U.S.C. §§ 2641-2656; 40 C.F.R. §§ 763.80-763.119; see also 59 Fed. Reg. 54,746 (Nov. 1, 1994) (proposing amendments to asbestos-in-schools regulations).

144/ See Iowa Code § 279.52; Iowa Admin. Code §§ 281-96.1-281.96.15. Under Iowa regulations, local educational agencies have extensive responsibilities.

145/ 49 U.S.C. § 1801; 49 C.F.R. §§ 172.101 (labeling requirements), 173.1090 (packaging requirements), 177.844 (minimization of exposure requirements).

146/ See 42 U.S.C. §§ 9601-9675; 40 C.F.R. § 302.4.

fire-resistant. PCBs generally includes polychlorinated biphenyl compounds and mixtures of substances that contain PCBs. "PCB items" are articles, containers, or equipment that contain or have been contaminated with PCBs.147/ Iowa regulations relating to PCBs are relatively limited.148/

Effective January 1, 1978, Congress banned with few exceptions the manufacture, processing, distribution, or use of any PCBs other than in a totally enclosed manner.149/ Under federal TSCA regulations, the manufacture, processing, and distribution of PCBs at concentrations of 50 parts per million (ppm) or greater is presumed to present an unreasonable risk of injury to health.150/ In general, provisions specifying a PCB concentration may not be avoided by dilution.

"Totally enclosed" activities include intact, non-leaking electrical equipment such as transformers and capacitors that contain PCBs in any concentration.151/ Certain exemptions may be obtained.152/ Other requirements apply to uses of PCBs in transformers (particularly in commercial buildings), electrical equipment, mining equipment, hydraulic systems, natural gas pipelines, small research projects, and other applications.153/

Marking and labeling (M_L) requirements apply to PCB-containing items and storage areas that contain PCB items,154/ including size and color requirements.155/

147/ 40 C.F.R. § 761.3.

148/ Iowa Admin. Code §§ 567-41.10(45) (drinking water supply standards for PCBs), 567-61.3 (surface water quality standards for PCBs), 567-100.1 and 121.3(4)(f) (restrictions on PCBs in solid waste), 567-119.2 (limitations on PCBs in waste oil).

149/ 15 U.S.C. § 2605(e)(2); 40 C.F.R. § 761.20(a).

150/ 40 C.F.R. § 761.20.

151/ 40 C.F.R. § 761.20.

152/ 40 C.F.R. § 761.20.

153/ 40 C.F.R. § 761.20.

154/ 40 C.F.R. § 761.40.

155/ 40 C.F.R. § 761.45.

The label depends on the type of equipment, the quantity and concentration of PCBs in the area, and other circumstances.156/ PCB records must be retained for three years in most circumstances.157/

As of December 1, 1985, all PCB transformers must be registered with local fire response personnel, and PCB transformers in use in or near commercial buildings must be registered with the building owner.158/ Other protection or removal requirements are applicable to different PCB transformers, including a prohibition on storage of combustibles in or within five meters of a transformer vault.159/ Because of the detailed requirements of these regulations, they should be consulted in connection with specific transformers.

In addition, TSCA regulations impose requirements on the disposal and storage of PCBs.160/ Although these requirements are similar to the RCRA hazardous waste storage and disposal requirements described in chapter 6 of this handbook, management of PCB waste differs slightly from other hazardous waste. Waste PCBs and PCB items can be stored for up to one year before disposal.161/ PCBs generally must be disposed of by incineration or by disposal in a chemical waste landfill.162/ Special federal regulations apply to incineration, chemical waste landfills, and decontamination of PCB items.163/ A manifest system tracks shipments of PCB waste.164/

156/ 40 C.F.R. § 761.40.

157/ 40 C.F.R. §§ 761.30(a)(1)(xii), 761.180(a).

158/ 40 C.F.R. § 761.30(a)(1)(vi).

159/ 40 C.F.R. § 761.30(a).

160/ 40 C.F.R. §§ 761.60-761.79; see also 40 C.F.R. §§ 761.202-761.218 (regarding PCB waste disposal records and reports).

161/ 40 C.F.R. § 761.65(a).

162/ 40 C.F.R. § 761.60.

163/ 40 C.F.R. §§ 761.70-761.79.

164/ 40 C.F.R. §§ 761.202-761.218.

For spills of PCBs in concentrations of 50 ppm or greater, reporting and sampling requirements apply.[165] In general, these requirements apply to spills that took place after 1978, and pre-1978 spills will be regulated on a site-by-site basis. Again, because of the detailed nature of these requirements, the relevant regulations should be consulted in connection with specific situations.[166] Finally, with regard to the cleanup of Superfund sites containing PCBs, the EPA issued a guidance document in August 1990. TSCA provides for civil and criminal penalties for violations of PCB regulations. Civil penalties of up to $25,000 may be assessed for each day that a violation continues.[167] Knowing and willful violations are subject to criminal punishment by a fine of not more than $25,000 for each day of violation, one year imprisonment, or both.[168]

13.11 Lead Regulation

Lead is another mineral that, like asbestos, was commonly used, but has the potential to cause human health effects. Lead has received increasing attention from regulators. In 1992, Congress amended TSCA by requiring EPA to develop a lead exposure reduction program.[169] The program is intended to reduce lead-based paint health hazards in connection with renovation and remodeling of homes constructed before 1978. While EPA adopts implementing regulations, EPA issued a guidance document dated July 14, 1994 providing advice on residential lead-based paint hazards including clearance levels considered unsafe by EPA.

165/ 40 C.F.R. § 761.120-761.135.

166/ 40 C.F.R. § 761.125; see also 56 Fed. Reg. 26,738 (June 10, 1991) (discussing proposed new rules).

167/ 15 U.S.C. § 2615(a); see also Guidelines for Assessment of Civil Penalties Under TSCA, 45 Fed. Reg. 59,770 (Sept. 10, 1980).

168/ 15 U.S.C. § 2615(b).

169/ 15 U.S.C. §§ 2681-2692; see also 59 Fed. Reg. 45,872 (Sept. 2, 1994) (to be codified at 40 C.F.R. pt. 745); 59 Fed. Reg. 49,484 (Sept. 28, 1994) (proposing EPA notification requirements prior to "significant new uses" of lead to be codified at 40 C.F.R. pt. 721).

Effective June 1988, amendments to the Safe Drinking Water Act banned the use of lead solder, pipes, or flux in drinking water systems by requiring "lead-free" materials.170/ Congress banned the sale of and required the recall of drinking water coolers containing lead.171/ Iowa has adopted maximum contaminant levels (MCLs) for lead in drinking water at Iowa Admin. Code § 567-41.4.

Pursuant to Clean Air Act provisions on regulation of fuels at 42 U.S.C. § 7545, the EPA regulates the level of lead in gasoline. These regulations have phased down the levels of lead in fuel and required lead-free fuel.172/

On July 14, 1994, EPA issued interim soil lead guidance for Superfund sites and RCRA corrective action facilities that generally recommend screening levels for lead in soil for residential use at 400 ppm as a trigger level for requiring further investigation. An earlier EPA Guidance Document, which apparently remains in effect, recommended soil lead cleanup levels of 500-1000 ppm at industrial Superfund sites.173/

In 1987, Iowa adopted legislation establishing a lead abatement program within the Iowa department of public health.174/ The program provides matching grants to municipalities for lead abatement programs. Iowa restricts land disposal of lead acid batteries, household batteries containing lead, and lead in packaging materials.175/

170/ 42 U.S.C. § 300g-6.

171/ 42 U.S.C. §§ 300j-22 and 23.

172/ 40 C.F.R. § 80.20 (0.10 grams of lead per gallon after January 1, 1986); see also Ethyl Corp. v. EPA, 541 F.2d 1 (D.C. Cir. 1976) (upholding phase-out regulations); Amoco Oil Co. v. EPA, 501 F.2d 722 (D.C. Cir. 1974) (upholding lead-free requirement).

173/ OSWER Dir. #9355.4-02 (Sept. 1989).

174/ Iowa Code §§ 135.100-135.105.

175/ Iowa Code §§ 455D.10-455D.10B, 455D.19.

CHAPTER FOURTEEN

ENVIRONMENTAL COMMON LAW AND TOXIC TORTS

14.0 Introduction

This chapter summarizes theories of common law liability relevant to environmental claims. It also highlights damages issues and statutes of limitations relating to environmental actions, including toxic tort claims. A "toxic tort" is personal injury or property damage caused by exposure to toxic or hazardous substances. Claims for toxic tort damages typically are asserted in a civil lawsuit alleging causes of action developed over time by the courts, known as the common law, such as nuisance, strict liability, trespass, and negligence, or a combination of these.

14.1 Nuisance

Traditionally, a nuisance action provides redress to persons injured by environmentally damaging activities. Iowa Code § 657.1 provides a cause of action as follows against a person who creates a nuisance:

> Whatever is injurious to health, indecent, or offensive to the senses, or an obstruction to the free use of property, so as to interfere with the comfortable enjoyment of life or property, is a nuisance, and a civil action by ordinary proceedings may be brought to enjoin and abate the same and to recover damages sustained on account thereof.

Although based on this statute, Iowa's nuisance law is largely a product of common law since Iowa courts have held that statutes defining nuisance do not supercede the common law of nuisance.1/ Courts have deemed a variety of

---

1/ Guzman v. Des Moines Hotel Partners, 489 N.W.2d 7 (Iowa 1992).

environmental injuries to be nuisances, including noise,[2] odors,[3] leakage of gasoline,[4] obstruction of streams,[5] and water pollution.[6] However, the potentially broad scope of the nuisance doctrine has been limited by the fact that determination of whether particular conduct is a nuisance requires a balancing of the gravity of harm against the utility of the defendant's conduct.[7]

A distinction is made between nuisances per se and nuisances in fact. A nuisance per se is an act which is a nuisance at all times and under all circumstances.[8] Iowa courts have declined to hold that refuse or sewage disposal facilities are nuisances per se though they may be operated so as to create a nuisance in fact.[9] The statute enumerates certain activities which will be deemed nuisances.[10] Otherwise, in determining whether a nuisance exists, the court considers factors such as the relative priority of the parties in relation to the location, the nature of the neighborhood, and the wrong complained of.[11] Further,

---

2/ See, e.g., Helmkamp v. Clark Ready Mix Co., 214 N.W.2d 126 (Iowa 1974); Bates v. Quality Ready-Mix Co., 261 Iowa 696, 154 N.W.2d 852 (1967); Schlotfelt v. Vinton Farmers' Supply Co., 252 Iowa 1102, 109 N.W.2d 695 (1961).

3/ See, e.g., Valasek v. Baer, 401 N.W.2d 33 (Iowa 1987); Patz v. Farmegg Prods., Inc., 196 N.W.2d 557 (Iowa 1972); Newton v. City of Grundy Center, 246 Iowa 916, 70 N.W.2d 162 (1955); Higgins v. Decorah Produce Co., 214 Iowa 276, 242 N.W. 109 (1932).

4/ See Mel Foster Co. Properties, Inc. v. American Oil Co. (Amoco), 427 N.W.2d 171 (Iowa 1988).

5/ See, e.g., Sioux City v. Simmons Warehouse Co., 151 Iowa 334, 129 N.W. 978 (1911), r'hng. denied and modified on other grounds, 151 Iowa 334, 131 N.W. 17 (1911).

6/ See, e.g., Newton v. City of Grundy Center, 246 Iowa 916, 70 N.W.2d 162 (1955); Perry v. Howe Co-op Creamery Co., 125 Iowa 415, 101 N.W. 150 (1904).

7/ Pitsenbarger v. Northern Natural Gas Co., 198 F. Supp. 665 (S.D.Iowa 1962).

8/ See, e.g., Ryan v. City of Emmetsburg, 4 N.W.2d 435 (1942); Stockdale v. Agrico Chem. Co., 340 F. Supp. 244 (N.D. Iowa 1972).

9/ See, e.g., Incorporated Town of Carter Lake v. Anderson Excavating & Wrecking, 241 N.W.2d 896 (Iowa 1976); Kriener v. Turkey Valley Comm. School Dist., 212 N.W.2d 526 (Iowa 1973).

10/ Iowa Code § 657.2.

11/ Helmkamp v. Clark Ready Mix Co., 214 N.W.2d 126 (Iowa 1974).

Iowa law holds that intent is irrelevant in determining whether an action constitutes a nuisance.12/

14.1.1 Private Nuisance and Farmer Protections

The term "private nuisance" refers to an actionable interference with a person's interest in private use and enjoyment of land.13/ To constitute a nuisance, one party's use of its property must interfere with another's use of adjoining property. The essential element of private nuisance is the injury to one's neighbor.14/ Iowa courts have defined a private nuisance as a civil wrong based on disturbance of rights in land including vibrations, blasting, destruction of crops, flooding, pollution, and disturbance of comfort as by unpleasant odors, smoke or dust.15/ In short, a private nuisance is an invasion of land that interferes with a neighboring individual's private use or enjoyment of such land.

Although plaintiffs sometimes have difficulty proving nuisance, the doctrine recognizes few defenses. A person responsible for a harmful condition found to be a nuisance may be liable even though that person has used a high degree of care to prevent or minimize the effect.16/

While no reported Iowa decision appears to have addressed the issue, the weight of authority tends to preclude use of a nuisance theory of recovery by a landowner against its predecessor in title.17/ The nuisance cause of action is

---

12/ See, e.g., Page County Appliance Center, Inc. v. Honeywell, Inc., 347 N.W.2d 171 (Iowa 1984); Kriener v. Turkey Valley Comm. School Dist., 212 N.W.2d 526, 530 (Iowa 1973); Patz v. Farmegg Prods., Inc., 196 N.W.2d 557, 561 (Iowa 1972).

13/ Bates v. Quality Ready-Mix Co., 261 Iowa 696, 154 N.W.2d 852 (1967).

14/ State v. Jacob Decker & Sons, 197 Iowa 41, 196 N.W. 600 (1924).

15/ Guzman v. Des Moines Hotel Partners, Ltd. Partnership, 489 N.W.2d 7 (Iowa 1992).

16/ Page County Appliance Center, Inc. v. Honeywell, Inc., 347 N.W.2d 171, 176 (Iowa 1984).

17/ See Allied Corp. v. Frola, 730 F. Supp. 626, 630 (D.N.J. 1990).

generally limited to interference with use of adjoining land.18/ Nevertheless, other states' courts have held that a plaintiff could sue in nuisance against the predecessor in title of adjoining land on which the nuisance arose, if that party was responsible for originating or continuing the harmful activity.19/

Some cases have held that a person who "comes to a nuisance" cannot bring a nuisance claim. The majority rule, however, is that a person who relocates into the vicinity of a nuisance may bring an action against the continued operation of a nuisance.20/ While Iowa follows the majority rule except as provided in Iowa Code § 172D.2, priority of occupation is "a circumstance of considerable weight" in determining whether such a suit will succeed.21/ Iowa Code § 172D.2, moreover, provides that in any nuisance action or proceeding against a feedlot brought by or on behalf of a person whose date of ownership of realty is subsequent to the established date of operation of that feedlot, proof of compliance with sections 172D.3 (state and federal environmental laws) and 172D.4 (county zoning) shall be an absolute defense, provided that the conditions or circumstances alleged to constitute a nuisance are subject to regulatory jurisdiction in accordance with either section 172D.3 or 172D.4.

In the Land Preservation and Use law, the Iowa legislature provided for the creation of "agricultural areas" subject to county approval.22/ As an incentive for agricultural land preservation, a farm or farm operation located in an agricultural area generally shall not be found to be a nuisance regardless of the established date

---

18/ See, e.g., Amland Properties Corp. v. Aluminum Co. of Am., 711 F. Supp. 784, 808 (D.N.J. 1989); Allied Corp. v. Frola, 730 F. Supp. 626, 630 (D.N.J. 1990).

19/ See, e.g., North Star Legal Found. v. Honeywell Project, 355 N.W.2d 186 (Minn.App. 1984).

20/ See generally 42 A.L.R. 3d 344 (1972); see also Spur Indus. v. Del E. Webb Dev. Co., 494 P.2d 700 (Ariz. 1972) (developer required to pay for cost of moving feedlot).

21/ Bates v. Quality Ready-Mix Co., 261 Iowa 696, 154 N.W.2d 852 (1968).

22/ Iowa Code § 352.6; see also Chapter 93A: Right-to-Farm Protection for Iowa, 35 Drake L. Rev. 633 (1985-1986).

of operation or expansion of the agricultural activities of the farm or farm operation.[23] This restriction on nuisance claims, however, does not apply to farm operations in violation of federal and state statutes and in certain other circumstances set forth in the statute. Under the same statute, a farmer that prevails in a nuisance lawsuit arising from a farm operation on farmland within an agricultural area may recover reasonable attorneys fees if the court determines that the claim is frivolous.

14.1.2 Public Nuisance

In addition to private nuisance actions, government agencies may bring actions for public nuisances. The same circumstances may create both a public and private nuisance.[24] The distinction between a public and private nuisance rests on whether the nuisance affects the rights of the public or the rights of an individual exclusively. A public or common nuisance is a species of catchall criminal offenses, consisting of an interference with the rights of a community at large.[25] Although individuals generally cannot initiate private actions for a purely public nuisance, some legal commentators have argued for elimination of the distinction between private and public nuisances to allow private individuals to bring such actions.[26]

Some courts, notably in California and New York, have recognized a cause of action based on prospective public nuisance, which allows plaintiffs to enjoin activity not yet producing any identifiable harm.[27] Reported Iowa case law does not appear to have addressed this theory.

23/ Iowa Code § 352.11.

24/ See Pottawattamie County v. Iowa Dept. of Environmental Quality, 272 N.W.2d 448 (Iowa 1978).

25/ Guzman v. Des Moines Hotel Partners, 489 N.W.2d 7 (Iowa 1992), quoting W. Page Keeton et al, Prosser and Keeton on the Law of Torts § 86, at 618-19 (5th ed. 1984).

26/ See Prosser, Private Action for Public Nuisance, 52 Va. L. Rev. 997 (1966).

27/ Helix Land Co., Inc. v. City of San Diego, 82 Cal. App. 3d 932, 147 Cal. Rptr. 683, 693 (1978).

14.2 Strict Liability

Under a strict liability theory, a person who keeps a potentially dangerous substance on his or her land which, if permitted to escape, is certain to injure others, must make good the damage caused by the escape of the substance regardless of the defendant's lack of negligence. This doctrine originated in the well-known British case of Rylands v. Fletcher.[28/] Iowa courts have "committed to a broader application of the strict liability doctrine of Rylands v. Fletcher than is reflected in the Restatement."[29/] Iowa does not limit strict liability to "ultrahazardous activity."

Plaintiffs often plead strict liability in an attempt to recover response costs for cleanup of hazardous substances.[30/] The Restatement of Torts suggests a case-by-case balancing of the existence of high risk, the likelihood that harm will result, the inability to eliminate the risk, and several other factors.[31/] In the context of hazardous waste cleanup disputes, the issue is whether disposal of hazardous waste constitutes abnormally dangerous activity, a question that reported Iowa case law has not yet resolved.[32/]

14.3 Negligence

A person who acts, or fails to act, when that person has a legal duty to do so is negligent if the action or inaction proximately causes another's injury. For example, an Iowa court has upheld a damage award based on negligence theories for persons

---

28/ 3 H.L. 330 (1868).

29/ National Steel Serv. Center v. Gibbons, 319 N.W.2d 269, 273 (Iowa 1982); see also Watson v. Mississippi River Power Co., 174 Iowa 23, 156 N.W. 188 (1916); Davis v. L. & W. Constr. Co., 176 N.W.2d 223, 224-25 (Iowa 1970).

30/ See Allied Corp. v. Frola, 730 F. Supp. 626, 630 (D.N.J. 1990) (predecessor landowner is absolutely liable for environmental damage to successor in title).

31/ See Restatement (2d) of Torts §§ 519-520 (1979).

32/ But see New Jersey v. Ventron Corp., 94 N.J. 473, 492-93, 468 A.2d 150, 160 (1983) (prior owners strictly liable for disposal of toxic wastes as abnormally dangerous activity).

harmed by defendants' mixing farm chemicals in the area of plaintiff's property.[33] Similarly, this theory may support recovery in a variety of other circumstances. A defendant's violation of a statute is prima facia evidence of negligence, provided that the action is of the kind which the statute was intended to prevent and the plaintiff is a member of the class which the statute is designed to protect.[34] Violation of government regulations or permit conditions, however, generally does not automatically establish negligence.[35]

14.4 Trespass

In the environmental context, trespass theories of liability typically apply to unlawful, forcible entry onto another's real property. The leading environmental case on trespass theory involved the deposit of microscopic fluoride compounds onto a neighbor's property from the vapor emitted by an aluminum processing plant.[36] Because this deposition of fluoride was a physical invasion of the property, the court held the defendant liable under a trespass theory.

Iowa cases have not explicitly considered trespass in conjunction with environmental liability, since environmental cases are often brought as nuisance actions. There may be cases, however, where trespass is more appropriate in that Iowa case law defines a trespass as "an actual physical invasion by tangible

---

33/ Kosmacek v. Farm Serv. Co-op of Persia, 485 N.W.2d 99 (Iowa App. 1992); see also Bloomquist v. Wapello County, 500 N.W.2d 1 (Iowa 1993).

34/ Wiersgalla v. Garrett, 486 N.W.2d 290, 292 (Iowa 1992); Koll v. Manatt's Transp. Co., 253 N.W.2d 265, 270 (Iowa 1977).

35/ See Lutz v. Chromatex, Inc., 29 Env't Rep. Cas. (BNA) 2045, 2057 (M.D. Pa. 1989); Brown v. Petrolane, Inc., 102 Cal. App. 3d 720, 162 Cal. Rptr. 551 (1980).

36/ See Martin v. Reynolds Metal Co., 221 Or. 86, 342 P.2d 790 (1959), cert. denied, 362 U.S. 918 (1960).

matter."37/ In some instances, an event may be both a nuisance and a trespass, and in the past, continuing trespasses have sometimes been dealt with as nuisances.38/

14.5 Inverse Condemnation

To be an "inverse condemnation," there must be a taking of property, but such a taking need not amount to a tranfer of physical control over assets.39/ Rather, plaintiff must merely show the loss of some compensable interest. Iowa courts have further held that the period of limitation for an inverse condemnation action is the general five-year period for actions brought for injury to property, not the ten-year period for actions brought for the recovery of real property.40/ Courts could well apply this theory to pollution or contamination related to governmental activity.

14.6 Infliction Of Emotional Distress

A recent Iowa appellate decision suggests that in some circumstances involving toxic torts, damages may be available for infliction of emotional distress.41/ Although the court in that particular case determined that the evidence did not support an award for mental suffering, it did acknowledge the possibility of such a remedy upon proper proof of requisite facts. The case involved claims against an adjoining landowner who, in negligently mixing farm chemicals, permitted the migration of herbicides onto plaintiff's property. The court held that

---

37/ Ryan v. City of Emmetsburg, 4 N.W.2d 435, 438 (Iowa 1942); see also Bethards v. Shivvers, Inc., 355 N.W.2d 39,47 (Iowa 1984) (trespass involving damage to farmland by reckless transportation of hay bales).

38/ Id. at 438-39; see also Newhall Land & Farming Co. v. Superior Court, 19 Cal. App. 4th 334, 23 Cal. Rptr. 2d 377 (1993) (property owner may bring claims of continuing nuisance, trespass, and negligence against former owners who allegedly contaminated property years ago).

39/ Water Dev. Co. v. Board of Water Works, 488 N.W.2d 158 (Iowa 1992).

40/ Scott v. City of Sioux City, 432 N.W.2d 144 (Iowa 1988).

41/ See Kosmacek v. Farm Serv. Co-op of Persia, 485 N.W.2d 99, 105 (Iowa App. 1992).

because there was no showing that plaintiffs were aware of an increased statistical likelihood of developing cancer, they could not have had a "reasonable apprehension which manifests itself as mental distress." The court reiterated that the emotional distress must be so severe that a reasonable man or woman could not be expected to endure it. In addition, the court noted that the mere possibilitity of future harm is insufficient -- there had to be a threshhold showing of reliable data linking the particular herbicide used by plaintiff with an increased risk of cancer.42/

A prima facie case of intentional infliction of emotional distress generally entails the following:

1. Outrageous conduct by the defendant;
2. The defendant's intention of causing, or reckless disregard of the probability of causing, emotional distress;
3. The plaintiff's suffering severe or extreme emotional distress; and
4. Actual and proximate causation of the emotional distress by the defendant's outrageous conduct.43/

While Iowa does not require physical injury to recover for emotional distress, Iowa courts disfavor recovery for mental anguish alone.44/

14.7 Battery

Personal injury claims related to toxic substance exposure may be based on the legal theory of battery. A person may recover in battery if someone disposes of toxic material with the intent to cause an offensive or harmful contact or knew that such

---

42/ Id. With regard to the admissibility of expert testimony in toxic tort cases, see Daubert v. Merrell Dow Pharm., 113 S. Ct. 2786 (1993); Bloomquist v. Wapello County, 500 N.W.2d 1 (Iowa 1993).

43/ See Amsden v. Grinnell Mut. Reinsurance Co., 203 N.W.2d 252, 255 (Iowa 1972); Meyer v. Nottger, 241 N.W.2d 911, 918 (Iowa 1976); Wambsgans v. Price, 274 N.W.2d 362, 365-66 (Iowa 1979).

44/ Wambsgans v. Price, 274 N.W.2d 362, 365 (Iowa 1979).

contact was substantially certain to occur.45/ Few cases, however, have addressed this issue in the environmental context.

14.8 Misrepresentation

Misrepresentation claims often arise in the environmental context in cases involving real estate purchases. Intentional misrepresentation, also referred to as fraud, involves misrepresentation of a material fact that induces another person to act. If a seller misrepresents to a buyer about a fact that materially affects the value of the property, this constitutes actionable fraud and may justify rescission of the contract. A misrepresentation need not be an affirmative misstatement and can arise from a failure to disclose material facts.46/

Generally, an action can be brought if the misrepresentation proximately causes injury that results in damages to the person who justifiably relied on the misrepresentation.47/ Some misrepresentations, however, may not reasonably be relied on.48/

14.9 Contribution, Indemnity, and Comparative Fault

A claim for contribution is a separate cause of action in Iowa. Contribution principles apply to situations where one party is required to pay more than its fair share of a common burden or obligation.49/ In environmental cleanup lawsuits, jointly liable parties commonly seek contribution from other parties who

---

45/ See, e.g., Werlein v. United States, 746 F. Supp. 887, 907 (D. Minn. 1990), vacated in part, 793 F. Supp. 898 (D. Minn. 1992).

46/ See Sinnard v. Roach, 414 N.W.2d 100, 105 (Iowa 1987).

47/ See Hall v.Wright, 261 Iowa 758, 766, 156 N.W.2d 661, 666 (1968).

48/ Lockard v. Carson, 287 N.W.2d 871, 878 (Iowa 1980).

49/ Restatement (Second) of Torts §§ 875-878. But see Beeck v. Aquaslide 'N Dive Corp., 350 N.W.2d 149, 170 (Iowa 1984) (reckless defendant barred from seeking contribution).

contributed to the contamination. Contribution claims are governed by the comparative fault act in Iowa Code § 668.7.

Indemnity also is a separate cause of action. Indemnity differs from contribution in that indemnity involves situations where one party is responsible for all of the damages rather than a comparative share. It remains in some question whether the comparative fault provisions in Iowa Code chapter 668 apply to indemnity claims.50/

With certain modifications, the uniform comparative fault act has been adopted in Iowa Code chapter 668 to govern tort disputes involving allocation of fault. Accordingly, contributory fault by a plaintiff is not a complete bar to recovery.51/ The comparative fault act contains numerous special provisions relating to allocation of fault, joint and several liability, the effects of releases, statutes of limitation, insurance coverage allocation, malpractice claims, products liability claims, and claims for interest.

14.10 Punitive Damages

The availability of punitive damages in tort cases is a controversial issue. Under Iowa law, punitive damages are incidental to the main cause of action.52/ There is no separate cause of action for punitive damages under Iowa law. Courts may award punitive damages in Iowa only when "a preponderance of clear and convincing evidence" reveals that the defendant's acts show wilful and wanton disregard for the rights or safety of another.53/ If the court or jury finds that the conduct of the defendant was specifically directed at the claimant, the claimant is

---

50/ American Trust & Savings Bank v. U.S. Fidelity, 439 N.W.2d 188, 190 (Iowa 1989).

51/ Iowa Code § 668.3.

52/ Burke v. Deere & Co., 6 F.3d 497, 507 (8th Cir. 1993), cert. denied, 114 S.Ct. 1063 (1994); see also Spaur v. Owens-Corning Fiberglas Corp., 510 N.W.2d 854 (Iowa 1994) (discussing punitive damages in asbestos case).

53/ Iowa Code § 668A.1.

entitled to the full amount of punitive damages awarded, but if the defendant's action was not specifically directed at the claimant, the claimant is only entitled to 25 percent of the the punitive damage award with the remainder paid into a civil reparations trust fund.[54/]

If a guardian, tenant for life or years, joint tenant, or tenant in common of real property commits "waste" on the property, that person is liable to pay three times the damages that have resulted from the waste to the person entitle to sue (usually the landowner).[55/] The concept of "waste" is somewhat similar to the nuisance principles discussed earlier in this chapter.

For willfully injuring any timber, tree, or shrub on the land of another, or in the street or highway in front of another's cultivated ground, yard, or city lot, or on the public grounds of any city or on any state land, the perpetrator is liable for treble damages.[56/] In a controversial case involving this statute, the Iowa Supreme Court reversed a district court's damage award for wrongfully destroyed trees, which was based on market value of lumber, and instructed the district court to take into account the sentimental, historic, and aesthetic value of trees on a Century Farm.[57/]

14.11 Municipal Liability

In Iowa, municipal liability is governed by Iowa Code chapter 670. Section 670.2 reads in part:

> Except as otherwise provided in this chapter, every municipality is subject to liability for its torts and those of its officers and employees,

---

54/ Iowa Code § 668A.1(2). With regard to constitutional restrictions on punitive damages awards, see Honda Motor Co. v. Oberg, 114 S. Ct. 2331 (1994); TXO Prod. Corp. v. Alliance Resources, Corp., 113 S. Ct. 2711 (1993); Pacific Mut. Life Ins. Co. v. Haslip, 499 U.S. 1 (1991).

55/ Iowa Code § 658.1.

56/ Iowa Code § 658.4.

57/ Bangert v. Osceola County, 456 N.W.2d 183, 190 (Iowa 1990); see also Hurley v. Youde, 503 N.W.2d 626 (Iowa App. 1993).

acting within the scope of their employment or duties, whether arising out of a governmental or proprietary function.

The act, which describes the tort liability of "governmental subdivisions," abrogates governmental immunity, except for certain exempted claims.58/ The act does not contain any express cap on liability, though punitive damages are not recoverable against a municipality.59/ Claims against a municipality or an officer, employee, or agent of a municipality must be brought within six months, unless the claimant notifies the governing body of the municipality within sixty days and states the time, place, and circumstances of the injury alleged, as well as the relief demanded, in which case the limitations period is extended to two years.60/

14.12 Iowa Tort Claims Act – Actions Against the State

The Iowa Tort Claims Act (ITCA) allows for the state appeal board established by Iowa Code § 73A.1 to consider and attempt to settle a claim before suit against the state is allowed.61/ No suit is permitted unless the state appeal board has made final disposition of a claim, or has failed to make final disposition of a claim within six months.62/ All claims and suits permitted by the ITCA are barred unless written notice of the claim is made to the state appeal board within two years after such claims accrued.63/

If compromise is not achieved, ITCA permits claims against the state for money only, "on account of damage to or loss of property or on account of personal injury or death, caused by the negligent or wrongful act or omission of any

---

58/ Iowa Code § 670.4 .

59/ See Iowa Code § 670.4(5).

60/ Iowa Code § 670.5.

61/ Iowa Code § 669.3.

62/ Iowa Code § 669.5.

63/ Iowa Code § 669.13; see also Vachon v. State, 514 N.W.2d 442 (Iowa 1994).

employee of the state acting within the scope of the employee's office or employment."[64] ITCA does not create any new cause of action but merely recognizes and provides remedy for a cause of action already existing, which would have been without remedy because of common law sovereign immunity.[65] Generally, the state is liable to the same extent as a private individual under like circumstances, except that it is not liable for interest prior to judgment or for punitive damages.[66] Also, there is a sizeable list of claims which are excepted from the provisions of the Act.[67]

14.13 Statutes of Limitations

Most of the relevant periods of limitation are assembled within Iowa Code § 614.1. Actions founded on injuries to person, whether based on contract or tort, must be brought within two years after the cause of action accrues.[68] Injuries to property and actions based on unwritten contracts, including actions for relief on the ground of fraud, must be brought within five years.[69] Actions based on written contracts, and those brought for recovery of real property, are viable for ten years.[70] In an action arising out of an unsafe condition of an improvement to real property based on tort and implied warranty, one may bring an action within 15 years.[71]

64/ Iowa Code § 669.2(3)(a).

65/ Engstrom v. State, 461 N.W.2d 309 (Iowa 1990).

66/ Iowa Code § 669.4 .

67/ Iowa Code § 669.14 .

68/ Iowa Code § 614.1(2); see also Sparks v. Metalcraft, Inc., 408 N.W.2d 347 (Iowa 1987).

69/ Iowa Code § 614.1(4); see also Hegg v. Hawkeye Tri-County REC, 512 N.W.2d 558, 560 (Iowa 1994) (applying 5-year statute of limitations in continuing negligence case).

70/ Iowa Code § 614.1(5).

71/ Iowa Code § 614.1(11).

CHAPTER FIFTEEN

INSURANCE

15.0 Introduction

Insurance coverage for environmental claims has been bitterly contested in Iowa and around the country. In some instances, insurance may be the only source of funding for environmental cleanups. On the other hand, insurance companies generally did not anticipate the retroactive environmental cleanup laws that were adopted in the 1980s and arguably did not take these types of costs into account in charging premiums.

Special insurance coverage may be purchased for environmental and pollution claims. With regard to insurance for underground storage tank response costs, chapter 10 of this handbook describes Iowa's insurance program. Most disputes, however, involve comprehensive general liability (CGL) insurance policies and umbrella insurance policies issued years ago. The standard CGL policy is an agreement under which the insurance company promises to pay on behalf of the insured all covered amounts that the insured becomes liable to pay others as damages. Generally, a claim is covered if it seeks recovery of "damages" for "property damage" to property owned by a person other than the insured or for "bodily injury" to a third-party, if the property damage or bodily injury took place during the policy period because of an "occurrence." This chapter discusses CGL coverage for environmental harm under Iowa law and general principles of Iowa insurance law that are relevant to claims under CGL policies.

15.1 Locating Insurance Policies

Under Iowa law, a party has the burden of proving each fact essential to the claim for relief.1/ Thus, the insured has the burden of proving the contract of insurance and its terms, as well as the loss. Accordingly, an insured generally has

---

1/ Dezsi v. Mutual Benefit Health & Acc. Ass'n, 255 Iowa 1027, 125 N.W.2d 219 (1963).

the obligation to locate the applicable insurance policies. The insurer must prove the contents of any exclusions or limitations on the amount of coverage on which it relies in defense.[2]

The best evidence rule codified in Iowa R. Evid. 1002 provides that, with some exceptions, no evidence other than the original of a writing is admissible to prove the contents of a writing. Under Iowa R. Evid. 1004, secondary evidence may be admitted to prove the contents of the writing if the original has been lost or destroyed unless the writing was lost or destroyed by the proponent of the writing "in bad faith" or with fraudulent intent. Once the insured has produced evidence showing that the policy has been lost or destroyed and that the insured has conducted a reasonably diligent but unsuccessful search for the missing policy, secondary evidence of the contents of the policy will be admissible.[3]

15.2 Choice Of Law In Insurance Disputes

An important step in analyzing a legal dispute over insurance coverage is determining which jurisdiction's law applies. In Wilmotte & Co. v. Rosenman Bros.,[4] the Iowa Supreme Court adopted the Restatement (Second) of Conflict of Laws to resolve choice of law questions concerning contracts. According to the Restatement (Second), the parties can designate which law is to control. In the absence of policy language stating which jurisdiction's law will apply, courts will apply the law of the state having the "most significant relationship" to the transaction in dispute.[5] Accordingly, Iowa law will generally, but not necessarily, apply to coverage disputes involving environmental contamination in Iowa.

2/ Bertran v. Glen Falls Ins. Co., 232 N.W.2d 527, 530 (Iowa 1975).

3/ Brintnall v. Professional Investors of Iowa, 218 N.W.2d 453, 456 (Iowa 1974); see also Cummins v. Pennsylvania Fire Ins. Co., 153 Iowa 579, 134 N.W. 79 (Iowa 1912).

4/ 258 N.W.2d 317 (Iowa 1977).

5/ Goetz v. Wells Ford Mercury, Inc., 405 N.W.2d 842, 843 (Iowa 1987); Cole v. State Auto. & Cas. Underwriters, 296 N.W.2d 779, 781 (Iowa 1980).

15.3 Iowa's Direct Action Statute

Iowa has adopted a form of direct action statute that allows an injured party to bring a direct action against the insurer of the party that caused the injury after obtaining judgment against that party and after that judgment has been returned as unsatisfied.6/ This statute gives an injured person an interest in a CGL policy adverse to both the insurer and the insured.7/

Although this statute does not authorize the injured party to add the insurer to the underlying action against the insured party, the injured party must bring the direct action against the insurer within 180 days from the entry of judgment against the insured.8/ Under Iowa's direct action statute, a person injured by environmental contamination may seek recovery under a responsible person's CGL policy if the responsible party cannot pay the judgment.

15.4 Occurrences

Under the typical CGL policy, coverage exists only for sums the insured becomes obligated to pay as damages because of bodily injury or property damage caused by an "occurrence." CGL policies usually define an "occurrence" as an "accident, including continuous or repeated exposure to conditions that result in bodily injury or property damage that was neither expected nor intended by the insured." In pollution cases, insurers often argue that there is no occurrence, and thus no coverage, because the insured must have expected that its activities, such as wastewater discharge or waste disposal, would cause environmental damage.

The Iowa Supreme Court has held that an accident, happening, event or exposure to conditions is an unexpected and unintentional "occurrence" so long as

---

6/ Iowa Code § 516.1.

7/ Opheim v. American Interins. Exch., 430 N.W.2d 118, 120 (Iowa 1988).

8/ Iowa Code § 516.3.

the insured does not expect or intend both the occurrence and some injury.[9] In West Bend Mut. v. Iowa Iron Works,[10] the insured deposited foundry sand on a nearby property. The Iowa DNR brought a suit against the foundry alleging the violation of sanitary disposal laws. The court held that the depositing of the sand on the nearby property, albeit intentional, was an occurrence because the foundry did not intend "some injury."[11]

15.5 Occurrences During the Policy Period -- Triggers of Insurance Coverage

CGL policies generally cover bodily injury or property damage resulting from an occurrence during the policy period. In a case involving bodily injury that manifests itself after a substantial latency period, which is not uncommon in toxic tort cases,[12] it can be difficult to establish when the injury occurred. Similarly, in a case involving property damage caused by migration of pollution in groundwater over an extended period of time, it can be difficult to establish when the damage was sustained. In First Newton Nat'l Bank v. General Cas. Co. of Wisconsin,[13] the Iowa Supreme Court held in a nonenvironmental context that the time of "occurrence" is when a claimant sustains actual damage and not when the act or omission causing that damage was committed.

15.6 Pollution Exclusions In Insurance Policies

Once an insured establishes that a claimed loss is within a CGL policy's general coverage provisions, the burden shifts to the insurer to show the

---

9/ First Newton Nat'l Bank v. General Cas. Co., 426 N.W.2d 618, 625 (Iowa 1988).

10/ 503 N.W.2d 596 (Iowa 1993).

11/ 503 N.W.2d at 601.

12/ Toxic tort actions are discussed in chapter 14 of this handbook.

13/ 426 N.W.2d 618, 624-25 (Iowa 1988).

applicability of any exclusion that will preclude coverage.14/ In addition, the insurer bears the burden of proof regarding an exception to any exclusion.15/ These cases often turn on the *contra proferentem* principle that when an insurance policy is ambiguous or is susceptible to two equally plausible conclusions, courts construe the policy against the drafter and apply the construction most favorable to the insured.16/

The so-called "standard" or "sudden and accidental" pollution exclusion clause, which has been the subject of litigation in many jurisdictions,17/ provides that no coverage exists for bodily injury or property damage "arising out of the discharge, dispersal, release or escape" of pollutants except if the discharge, dispersal, release or escape is "sudden and accidental." Generally, pre-1973 CGL policies did not contain a pollution exclusion, and policies issued between 1973 and 1985 contained the sudden and accidental pollution exclusion. CGL policies issued after 1985 usually contain a so-called "absolute" pollution exclusion that does not have an exception for "sudden and accidental" releases.

The most significant Iowa decision to address the application of the pollution exclusion with the "sudden and accidental" exception is Weber v. IMT Insurance Co.18/ In that case, the Webers operated a farm raising hogs and crops, using hog manure to fertilize the crops. In Weber, hog manure from the insured's operation repeatedly spilled on the road. The Iowa Supreme Court held that the Webers knew

---

14/ A.Y. McDonald Indus. v. INA, 842 F. Supp. 1166, 1171 (N.D. Iowa 1993); Kalell v. Mutual Fin. & Auto. Ins. Co., 471 N.W.2d 865, 867 (Iowa 1991).

15/ A.Y. McDonald, 842 F. Supp. at 1171.

16/ West Bend Mut. v. Iowa Iron Workers, 503 N.W.2d 596, 598 (Iowa 1993); Benzer v. Iowa Mut. Tornado Ins. Ass'n, 216 N.W.2d 385, 388 (Iowa 1974).

17/ Compare Just v. Land Reclamation Ltd., 155 Wis.2d 737, 456 N.W.2d 570 (Wis. 1990) (sudden and accidental exclusion is ambiguous and does not exclude coverage) with Sylvestor Bros. Dev. Co. v. Great Cent. Ins. Co., 480 N.W.2d 368 (Minn. App. 1992) (sudden and accidental exclusion is unambiguous and precludes coverage).

18/ 462 N.W.2d 283 (Iowa 1990).

or should have known that manure was going to spill on the road when it was transported in their manure spreaders. According to the court, the district court's factual finding that the spills were expected by the Webers and therefore not "accidental" was supported by substantial evidence. Because the spills were not accidental, the court held that the "sudden and accidental" exception to the pollution exclusion clause did not apply. The Weber court's "accidental" analysis focused on the discharge of the hog manure -- not the resulting damage caused by the discharge. Thus, under Iowa law, the pollution exclusion looks at the discharge, whereas the occurrence clause looks at the resulting damage caused by the discharge.

The federal district court in A.Y. McDonald[19] explained the distinction:

> [U]nder the accidental prong of the "sudden and accidental" exception to the pollution exclusion, a court in Iowa should ask: was the discharge expected or intended? If the discharge was either expected or intended, the discharge could not be accidental, and the pollution exclusion would preclude coverage. Conversely, in determining whether an occurrence has occurred, an Iowa court would ask: was the resultant damage caused by the discharge expected or intended?

15.7 Damages Versus Equitable Relief

The insuring clause in CGL policies usually states that the insurer will pay, on behalf of the insured, "damages," imposed on the insured by law. In many environmental cases, the insured is compelled to bear the costs of cleaning up pollution through means other than a lawsuit seeking "damages," such as a suit seeking injunctive relief or an administrative order requiring a cleanup. In other environmental cases, the insured is obligated to reimburse the government or a private party for their response costs, which insurers may argue is the equitable remedy of restitution.[20] In 1991, the Iowa Supreme Court ruled that the term

---

19/ 842 F. Supp 1146, 1174 (N.D. Iowa 1993).

20/ Compare City of Edgerton v. General Cas. Co., 184 Wis.2d 750, 517 N.W.2d 463 (Wis. 1994) (cleanup costs are equitable costs and not "damages") with Minnesota Mining & Mfg. Co. v. Travelers Indem. Co., 457 N.W.2d 175, 179-81 (Minn. 1990) (cleanup costs are "damages").

"damages" in the CGL policies at issue in that case includes government-mandated response costs under CERCLA and similar state environmental statutes, but not any civil penalties imposed thereunder.21/

15.8 Property Damage

Even if a claim seeks to recover "damages" from an insured, CGL coverage exists only if the claim for "damages" seeks recovery for bodily injury or property damage. CGL policies usually define property damage as "injury to or destruction of tangible property." Insurers may argue that cleanup costs are economic losses, regulatory compliance costs, or remedial costs imposed by law, but are not amounts payable on account of property damage.

In Kartridg Pak v. Travelers Indem. Co.,22/ the underlying claimant asserted that a mechanical deboner it leased from the defendant did not perform as promised, thereby causing the failure of its business. In the dispute with its insurer, the defendant contended that the failure of the deboner to sufficiently separate the meat and bone after grinding up the backbones, and thereby diminishing their value, constituted "property damage" triggering insurance coverage. The court focused on the policy language that required "physical injury to or destruction of tangible property" and not mere "injury to tangible property." The court concluded that intangible damages, such as diminution in value, did not constitute physical injury to or destruction of tangible property. Thus, the insurer's duty to defend was not triggered.

However, in First Newton Nat'l Bank v. General Cas. Co. of Wisconsin,23/ the Iowa Supreme Court held that the definition of "property damage" in the policies did not require physical injury to tangible property before loss of use was covered.

21/ A.Y. McDonald Indus., Inc. v. Insurance Co. of N. Am., 475 N.W.2d 607, 620, 626 (Iowa 1991).

22/ 425 N.W.2d 687 (Iowa App. 1988).

23/ 426 N.W.2d 618, 624-26 (Iowa 1988).

The court reasoned that tangible property rendered useless is injured and hence covered and that the term "property damage" did not require actual physical damage but could include tangible damage such as diminution in value of tangible property.

In 1991,[24] the Iowa Supreme Court held that any injury to the environment that results from hazardous waste contamination constitutes "property damage" within the meaning of CGL policies. Yet, on the facts of that case, the court found the record insufficient to determine whether the remedial measures would be necessary "because of property damage." The court held that response costs for preventive measures employed after pollution has occurred are incurred "because of property damage." The court ruled, however, that, in the absence of property damage, preventive measures are not covered under CGL policies. The court also noted that costs of compliance with governmental safety regulations are not incurred because of property damage, but instead are incurred to insure safety.

15.9 Notification Of Insurers

A CGL policy generally requires the insured to give the insurer timely notice of a lawsuit, a claim against it, or an occurrence. The purpose of the notice requirement is to enable insurers to make timely and thorough investigations of claims filed under their policies. The general rule in Iowa is that failure to provide timely notice will relieve an insurer of its obligations under a policy that expressly requires timely notice.

In Iowa, there is a rebuttable presumption of prejudice to the insurer due to delayed notice.[25] However, where there exists some legal justification or excuse for the delay, then the defendant insurer must demonstrate actual prejudice. Even if

---

24/ A.Y. McDonald Indus., Inc. v. Insurance Co. of N. Am., 475 N.W.2d 607, 620, 626 (Iowa 1991).

25/ Estate of Wade v. Continental Ins. Co., 514 F.2d 304 (8th Cir. 1975); Henderson v. Hawkeye-Security Ins. Co., 252 Iowa 97, 106 N.W.2d 86 (1960).

the delay is not excused, the insured may still recover if the insured can demonstrate that the insurer suffered no actual prejudice from the delay.26/

Although it should be apparent to an insured when a lawsuit has been brought against it that triggers its duty to notify, it may be difficult to determine whether there has been an occurrence or claim that triggers the duty. For example, if a company learns that in the past it sent hazardous waste to a site that currently has significant soil or groundwater contamination, it may decide that there is an occurrence because the company's actions contributed to property damage that was unexpected and unintended. Nevertheless, the company may have difficulty deciding whether that occurrence is likely to give rise to a claim for damages. In general, it is prudent for a company to notify its insurer whenever it has reason to believe that property damage or bodily injury has occurred that might be attributable to the company's waste disposal activities.

15.10 Duty To Defend Against "Suits"

Under a CGL policy, the insurer usually has a duty to defend the insured as well as to pay the insured's ultimate liability. The duty to defend is distinct from and broader than the duty to indemnify. The existence of the duty to defend depends on the allegations made in the complaint against the insured. If the complaint contains any allegations that are actually or potentially within the scope of coverage afforded by the policy, the insurer must defend.27/ The Iowa Supreme Court has explained that an insurer must defend an environmental suit against an insured "whenever there is potential or possible liability to indemnify the insured based on facts appearing at the outset of the case."28/ Thus, although coverage for

---

26/ Estate of Wade v. Continental Ins. Co., 514 F.2d 304 (8th Cir. 1975); Henschel v. Hawkeye-Security Ins. Co., 178 N.W.2d 409, 415 (Iowa 1970).

27/ First Newton Nat'l Bank v. General Cas. Co. of Wisconsin, 426 N.W.2d 618, 629-30 (Iowa 1988).

28/ A.Y. McDonald Indus., Inc. v. Insurance Co. of N. Am., 475 N.W.2d 607, 627 (Iowa 1991).

cleanup costs imposed on an insured may ultimately not exist, the insurer may be required to pay the insured's costs of defense.29/

The primary insurer is not relieved of its exclusive duty to defend until its limits of liability have been exhausted by settlement or judgment, not by payment of defense costs. As a result, an excess insurer generally owes no duty of defense to a common insured until all underlying limits of liability have been exhausted.30/

Finally, the Iowa Supreme Court has ruled that EPA proceedings constituted a "suit" within the meaning of CGL policies sufficient to trigger an insurer's duty to defend, even though no action had been filed in a court of law.31/ The court found the term "suit," which was undefined in the policies, to be ambiguous. Consequently, the court adopted the broader meaning of the term and defined it to include "any attempt to gain an end by legal process."

15.11 Consequences Of Failure To Defend

When an insurer fails to defend its insured because it has determined that there is no potential or possibility of liability, it acts "at its own risk."32/ By refusing to defend, the insurer waives its right to employ counsel to represent the insured to otherwise control the defense of the third-party action against the insured.33/

A failure to defend also will result in the insurer being liable for all damages reasonably incurred by the insured. If there was coverage -- or even potential coverage -- the wrongful failure to defend may be held to be a breach of the insurer's

---

29/ West Bend Mut. v. Iowa Iron Workers, 503 N.W.2d 596, 601 (Iowa 1993).

30/ But see Farm & City Ins. Co. v. United States Fid. & Guar. Co., 323 N.W.2d 259 (Iowa 1982) (where the primary insurer establishes that the excess carrier would have participated in the defense a pro rata distribution of litigation expenses may be ordered).

31/ A. Y. McDonald Industries, Inc. v. Insurance Co. of N. Am., 475 N.W.2d 607, 627-29 (Iowa 1991).

32/ Priester v. Vigilante Ins. Co., 268 F. Supp. 156 (S.D. Iowa 1967); Dairyland Ins. Co. v. Hawkins, 292 F. Supp. 947 (S.D. Iowa 1968).

33/ Kooyman v. Farm Bureau Mut. Ins. Co., 315 N.W.2d 30 (Iowa 1982).

covenant of good faith and fair dealing with the insured. If so, the insurer risks extracontractual liability as well as breach of contract liability. Finally, if the insurer's failure to defend was in "conscious disregard" of the insured's rights, punitive damages may be imposed.34/

34/ Pirkl v. Northwestern Mut. Ins. Assn., 348 N.W.2d 633 (Iowa 1984); see also Kooyman v. Farm Bureau Mut. Ins. Co., 315 N.W.2d 30 (Iowa 1982) (insurance contract includes implied covenant of good faith and fair dealing); Iowa Code § 507B.4 (enumerating unfair or deceptive acts or practices in insurance business).

CHAPTER SIXTEEN

ENVIRONMENTAL CONSIDERATIONS IN REAL ESTATE AND BUSINESS TRANSACTIONS

16.0 Overview

Parties involved in transactions that transfer an interest in real estate may find themselves, as owners or operators of the property, strictly as well as jointly and severally liable for environmental cleanup costs even though they had no connection with the disposal activity, subject to the innocent landowner exclusion discussed below.[1/] Moreover, the burden of demonstrating so–called "innocence" may be difficult. Accordingly, current owners and operators of property should not rely upon the availability of this statutory exemption to liability as their sole protection from liability.

One of the most important components of advance planning for parties involved in real estate transactions and business transactions in which real property is transferred is for the parties to retain an experienced environmental attorney at the outset of negotiations. The attorney should have a thorough understanding of the intricate environmental laws and regulations and be able to recognize the environmental pitfalls that may not be apparent to non–environmental specialists. An environmental attorney can supervise the site investigation to apprise the parties of the true condition of the property and can draft protective language in the transactional documents during the negotiations rather than performing limited "damage control" after most of the terms of the transaction have already been negotiated.

16.1 Transferor's Duty to Disclose Contamination; "As Is" Transfers

Iowa Code § 558.69 provides that the transferor of real property must prepare and file an environmental disclosure statement with the county recorder. The transferor must state that "no known wells are situated on the property," that the

---

1/ See chapter 5 of this handbook for a discussion of liability under Superfund laws.

property is not a "known disposal site for solid waste," "no known underground storage tanks" exist on the property, and that "no known hazardous waste" exists on the property. If wells or tanks exist on the property, the statement must provide information regarding the wells or tanks. If hazardous waste exists on the property, the transferor must attest that the waste is being managed in accordance with applicable regulatory requirements. The term "exists" is not defined, but it presumably would include hazardous waste stored in containers or released into the ground.

Iowa Code § 455B.430 requires approval from the DNR Director before the sale, conveyance, or transfer of or substantial change in use of hazardous waste sites which are on Iowa's registry of hazardous waste sites. Properties may be on the registry if the property was used for the disposal of hazardous waste or substances either illegally or prior to hazardous waste regulation or if the property adjoined a hazardous waste site and was affected by the disposal activities. Iowa's statutes require the DNR to file with the county recorder a statement disclosing the period during which all sites placed on the state registry were used as a waste disposal area.[2]

Effective July 1, 1994, any person wishing to transfer residential real property with "at least one but not more than four dwelling units" must provide the transferee with a disclosure statement identifying the "condition" and "important characteristics" of the property prior to accepting a written offer for the transfer of the real property.[3] Important characteristics would most likely include known environmental defects.[4] Persons violating this provision are subject to liability for all damages sustained.[5]

2/ Iowa Code § 455B.431.

3/ Iowa Code §§ 558A.1, 558A.2.

4/ Iowa Code § 558A.4.

5/ Iowa Code § 558A.5.

A seller may be liable for misrepresenting to a prospective buyer, by action or inaction, the condition of real property. For example, where a landowner discovered mercury contamination of its property caused by prior owners and intentionally failed to disclose this contamination to a buyer, the landowner was held liable to the subsequent purchaser for costs to remedy the contamination and, potentially, for compensatory damages such as diminution in value of the property.[6/]

Similarly, where a seller knew that contamination existed and told an interested buyer that past problems had been remedied, a statement the seller knew was false when made, the court found fraud. Even "as is" and "no representations and warranties" clauses in the contract could not protect the seller from the misrepresented material facts.[7/] However, many courts have been reluctant to disregard such "as is" clauses or similar provisions when the parties are of relatively equal bargaining strength and both have some financial sophistication.[8/]

Generally, language in contracts relating to the sale of real estate, where unambiguous, is controlling. Although the doctrine of caveat emptor or "let the buyer beware" has been eroded in many states in recent years, courts appear willing to apply the doctrine in cases involving the sale of commercial and industrial realty between two sophisticated corporations, reasoning that the purchasers should have the expertise to determine the condition of the property prior to sale.[9/] However,

---

6/ State v. Ventron Corp., 94 N.J. 473, 468 A.2d 150, 166–67 (1983); see also T&E Indus., Inc. v. Safety, 587 A.2d 1249 (N.J. 1991); Newson v. Reichold Chemicals, Inc., No. H–86–0077(G) (D. Miss. 1988), 3 Toxics Law Rptr. 519 (Sept. 21, 1988).

7/ Gopher Oil Co. v. Union Oil Co., 955 F.2d 519 (8th Cir. 1992); Ocean Cape Hotel Corp. v. Masefield Corp., 63 N.J. Super. 369 (App. Div. 1960); see also Southland Corp. v. Ashland Oil, Inc., 696 F. Supp. 994 , 1001 (D.N.J. 1988) ("as is" provision precluded only warranty claims, not contribution claim for cleanup costs under CERCLA, 42 U.S.C. § 9613(f)).

8/ See, e.g., La Placita Partners v. Northwestern Mut. Life Ins. Co., 935 F.2d 270 (6th Cir. 1991); Smith Land & Improvement Corp. v. Rapid-American Corp., 26 Env't Rep. Cas. (BNA) 2,023 (M.D. Pa. 1987); In re Hemingway Transport, Inc., 73 Bankr. R. 494 (Bankr. D. Mass. 1987).

9/ See, e.g., Philadelphia Electric Co. v. Hercules, Inc., 762 F.2d 303, 312-13 (3rd Cir.), cert. denied, 106 S.Ct. 384 (1985).

the doctrine of caveat emptor does not defeat a claim by a purchaser of property for contribution or indemnity from the seller for cleanup costs under CERCLA.[10/]

In addition, CERCLA's "innocent landowner" protection, discussed in section 16.3, is not available to any property owner who transfers property without disclosing actual knowledge of contamination obtained during the time it owned the property.[11/]

16.1.1 Rescission

Under Iowa law, a party may rescind a contract if it was induced to enter the contract by either a fraudulent or material misrepresentation.[12/] A misrepresentation is fraudulent if it is intended to induce formation of a contract and is known to be false or made without knowledge of whether it is true or false. If rescission is based on fraud, evidence of fraud must be clear.

A party, however, need not show fraudulent intent if an innocent misrepresentation is "material." A misrepresentation is material if it would be likely to induce a reasonable person to assent or if the person making the misrepresentation has special reason to know the misrepresentation will cause the recipient to assent. Rescission based on misrepresentation is closely related to rescission based on breach of contract. A material breach of a contract or a substantial failure in its performance justifies the other party in rescinding.[13/]

Under Iowa law, rescission of a contract is also warranted based on a mutual mistake. A mutual mistake is a clear misunderstanding that is common to both

---

10/ Smith Land & Improvement Co. v. Celotex Corp., 851 F.2d 86 (3d Cir. 1988), cert. denied, 488 U.S. 1029 (1989); FMC Corp. v. Northern Pump Co., 668 F. Supp. 1285, 1292 (D. Minn. 1987).

11/ 42 U.S.C. § 9601(35)(C).

12/ Maytag Co. v. Alward, 253 Iowa 455, 112 N.W.2d 654 (Iowa 1962); Folkers v. Southwest Leasing, 431 N.W.2d 177, 181 (Iowa App. 1988); see also Restatement of Contracts (Second) § 164(1) (1981).

13/ See Lutz v. Cunningham, 240 Iowa 1037, 38 N.W.2d 638 (Iowa 1949); Binkholder v. Carpenter, 260 Iowa 1297, 152 N.W.2d 593 (Iowa 1967); see also Distronics Corp. v. Roberts–Hamilton Co., 575 F. Supp. 275, 277 (D. Minn. 1983).

parties and concerns the terms and subject matter of a contract, or some substantial part of it.[14]

Although reported Iowa case law apparently has not addressed rescission of a real estate purchase agreement based on the discovery of environmental contamination, courts in other jurisdictions have. These courts have applied the same general principles reflected in Iowa law as discussed above. For example, in Continental Concrete Pipe Corp. v. K & K Sand & Gravel, Inc.,[15] a buyer of property contaminated with hazardous substances sought to rescind the agreements by which it acquired the property. The court held that both fraud and mutual mistake relating to environmental contamination were grounds for setting aside a contract.

In Garb–Ko, Inc. v. Lansing–Lewis Servs., Inc.,[16] a seller sought to rescind a real estate contract after the buyer discovered environmental contamination. The court concluded that, in entering the contract, the parties had made a basic material assumption that the property was not contaminated. The court held that the contract's "as is" clause did not affect the mutual mistake made by the parties and upheld rescission of the contract.[17]

14/ Folkers v. Southwest Leasing, 431 N.W.2d 177, 181 (Iowa App. 1988); see also Dubbe v. Lano Equip., Inc., 362 N.W.2d 353, 356 (Minn. App. 1985).

15/ 1990 Westlaw 7,095 (N.D. Ill. Jan. 22, 1990).

16/ 167 Mich. App. 779, 423 N.W.2d 355, 356 (1988).

17/ See also United States v. Fred Webb Inc., No. 89–41–2, 4 Toxics L. Rep. 979 (E.D.N.C. Jan. 31, 1990) (discussing argument in unresolved case that because seller fraudulently concealed environmental contamination, contract is void); Newsom v. Reichhold Chems. Inc., No. H86-0077(G), 3 Toxics L. Rep. 519 (S.D. Miss. Sept. 21, 1988) (discussing settlement of case involving rescission of real estate contract and imposition of damages based on fraudulent concealment of environmental contamination; Albanese v. City Fed. Sav. & Loan Ass'n, 710 F. Supp. 563, 564–65 (D.N.J. 1989) (noting buyer's viable action for fraud based on concealment of environmental contamination in proceeding concerning RICO liability; Levin Metals Corp. v. Parr–Richmond Terminal Co., 608 F. Supp. 1272, 1276–77 (N.D. Cal. 1985) (remanding cause of action for fraudulent concealment of environmental contamination to state court), aff'd and rev'd on other grounds, 799 F.2d 1312, 1315 (9th Cir. 1986).

16.2 Environmental Lien Provisions

CERCLA provides a federal lien for response costs or damages.[18] The lien may be imposed against all property belonging to a responsible person that is subject to or affected by removal or remedial action.[19] Priority is determined by time of filing of notice of the federal lien. The lien is specifically subject to the rights of any purchaser, holder of a security interest, or judgment lien creditor whose interest is perfected under applicable state law before the notice of lien is filed.[20]

16.3 Innocent Landowner Protection

The 1986 SARA amendments to CERCLA impose an affirmative obligation on acquiring parties to undertake "all appropriate inquiry into the previous ownership and uses of the property consistent with good commercial or customary practice,"[21] so that their innocence is bona fide. Ostriches will not be held to be innocent owners. Thus, the prudent acquirer of real property (whether by purchase, merger, acquisition, or foreclosure) will undertake a "due diligence" investigation to ascertain the condition of the property and thereby minimize potential liability. The "catch 22" is, of course, that the more diligent the inquiry, the less the chance that the transferee will be "innocent" when it acquires its interest in the property.

The requisite level of inquiry will necessarily vary from case to case. In assessing whether the duty to inquire has been satisfied, SARA directs the courts to take five factors into account:

1. Any specialized knowledge on the owner's part;

18/ 42 U.S.C. § 9607(l), (m).

19/ These terms are defined and discussed further in chapter 5 of this handbook.

20/ 42 U.S.C. § 9607(l)(3).

21/ 42 U.S.C. § 9601(35)(B); United States v. Pacific Hide & Fur Depot, 30 Env't Rep. Cas. (BNA) 1082 (D. Idaho 1989); United States v. Serafini, 28 Env't Rep. Cas. (BNA) 1162, 1167–68 (M.D. Pa. 1988).

2. The relationship of the purchase price to the value of the property if uncontaminated;

3. Commonly known or reasonably ascertainable information about the property;

4. The obviousness of the presence of likely contamination at the property; and

5. The ability to detect such contamination by appropriate inspection.[22]

Although these factors are designed to apply to CERCLA, they provide guidance for any purchaser to determine the prudent level of inquiry required.

On March 15, 1993, ASTM (formerly the American Society for Testing and Materials), a private organization that develops voluntary consensus standards, passed two standards designed to define the level of investigation necessary to satisfy the preacquisition inquiry element of CERCLA's innocent owner defense.[23] Since the language of CERCLA calls for inquiry consistent with "good commercial or customary practice" and the ASTM standards grew out of consensus of a committee comprised of representatives from the fields of real estate development, commercial lending, environmental consulting, industry, government, and academia, the ASTM standards provide guidance as to the level of inquiry appropriate to qualify for the innocent owner defense.

In addition to investigating the property, to qualify for the defense, CERCLA requires an innocent purchaser to establish that it exercised due care with respect to the hazardous substances and took precautions against foreseeable acts or omissions of third parties and the consequences of such acts or omissions.[24] As discussed in chapter 5 of this handbook, these provisions have not been meaningfully interpreted by the courts.

22/ 42 U.S.C. §§ 9601(35), 9607(b)(3).

23/ ASTM Standard Practice For Environmental Site Assessments: Standard E 1528-93 Transaction Screen Process; Standard E 1527-93 Phase I Environmental Site Assessment Process.

24/ 42 U.S.C. §§ 9607(b)(3), 9601(35)(A).

16.4 Environmental Site Assessments

Whether or not the person acquiring real property seeks to establish "innocent landowner" protection, an environmental site assessment is necessary to apprise the parties of the site's true condition, and thus the potential liability of the transferee. The purpose of the assessment is to identify all known or suspected environmental hazards associated with the property and to provide enough information about those hazards to enable those relying on the report to evaluate their exposure and take steps to manage or allocate risk. Environmental assessments are generally conducted in phases, although with sites known or strongly suspected to be contaminated, a single comprehensive investigation may save time and money.

A so–called "Phase I" audit provides a general overview of the site. It includes a visual inspection of the property and any improvements located on it and an investigation into past and present ownership and uses. It also includes a review of regulatory agency files and databases related to the site and adjacent properties. A "Phase II" assessment typically involves a field investigation consisting of soil borings and the installation of monitoring wells to ascertain the condition of groundwater beneath the site and, hopefully, to identify the source of any contamination discovered.

The ASTM standards introduced the possibility of an inquiry less rigorous than a typical Phase I. ASTM's "Transaction Screen," which may be conducted by either a lay person or an environmental professional, involves (1) a checklist of questions to ask the owner or operator of the property; (2) a site visit; and (3) a records review. The Transaction Screen is intended for transactions too small to bear professional fees and which, in the past, may have been consummated without any environmental inquiry whatsoever. If a Transaction Screen raises suspicions of contamination, the investigating party is directed to go on to a Phase I.

As noted in the previous section, CERCLA requires a would–be innocent landowner to demonstrate by a preponderance of the evidence that "all appropriate inquiry into the previous ownership and uses of the property consistent with good commercial or customary practice in an effort to minimize liability" was undertaken.[25] Beyond the Transaction Screen introduced by the ASTM Standards, "good commercial or customary practice" requires that in commercial transactions the purchaser retain competent professionals to conduct the site investigation.[26] Environmental legal counsel and an environmental consultant should be retained at an early stage. To maximize protection of the data developed under the attorney-client privilege and attorney work product doctrines, the environmental attorney, rather than one or both of the parties, should retain the environmental consultant. Preferably the consultant would be hired pursuant to a retainer letter clearly spelling out the purpose and scope of the investigation and providing for the delivery of all reports in draft form to the attorney for review. Factual misstatements, inadequate or irrelevant technical investigations, and erroneous legal advice are not uncommon in assessment reports prepared by some technical consultants. Unfortunately, once prepared, these reports have come back to haunt many a party in a real estate transfer.

Although the inquiry will vary somewhat from case to case, the following discussion and checklists suggest areas and sources of information that should be considered by any party acquiring real property or financing such an acquisition.

25/ 42 U.S.C. § 9601(35)(B); see also EPA Guidance on Superfund Program: De Minimis Settlements, Prospective Purchaser Settlements, 54 Fed. Reg. 34235, 34238 (Aug. 18, 1989); United States v. Serafini, 28 Env't Rep. Cas. (BNA) 1162, 1167–68 (M.D. Pa. 1988).

26/ The legislative history of this section indicates that "those engaged in commercial transactions should . . . be held to a higher standard than those who are engaged in private residential transactions." H.R. Rep. No. 99–962, 99th Cong. 2d Sess. 187 (1986).

16.4.1 On–Site Inspection

The site should be inspected by a qualified consultant, preferably in the company of environmental legal counsel, to look for obvious environmental problems. This is also a good opportunity for the attorney to interview knowledgeable employees or officers to elicit information regarding past ownership and uses. The walk–through inspector will be looking for hazardous materials such as asbestos, PCB-containing electrical equipment, urea, formaldehyde, and for signs that the property is or may be contaminated by hazardous substances. The inspection should include, to the extent feasible, reconnaissance of adjoining properties to determine whether it is likely that off–site contamination may be migrating onto the property.

16.4.2 Ownership and Operational History

The environmental consultant should conduct a thorough review of the history of the site and its prior uses. Some of this information will be gathered through the on–site physical inspection of the property. In addition, the consultant should inspect relevant company records and obtain copies of all operating permits, reports filed with government agencies, and hazardous waste disposal records. Interviews with key present and former officers and operational employees are critical. The consultant should determine at least the following:

1. Present and prior owners;
2. Present and prior uses;
3. Standard Industrial Classification (SIC) code numbers for present and prior users;
4. Site topography, including alterations such as filling or excavation;
5. Raw materials used;
6. Method of delivery and storage of hazardous materials;
7. History of former and existing underground tanks (contents, age, integrity);
8. Present and prior processes including waste streams related to each;
9. Location of storage areas, storage vessels, and surface impoundments;
10. On–site and off–site waste disposal methods;

11. Government permits and approvals necessary for operations, expiration dates, and transferability of permits;
12. Government inspections, notices of violations, and remedial activities undertaken;
13. Presence of asbestos;
14. Presence of PCB electrical equipment;
15. Location and use of on–site wells;
16. Septic systems, wastewater treatment facilities, and sump pits (location and points of discharge);
17. Storm water runoff practices;
18. Right–to–know postings and filings;
19. Environmental and CGL insurance applications, policies, claims and inspections;
20. Financial reserves for environmental liabilities;
21. Pollution control budgets and reports of capital expenditures for pollution control;
22. Waste oil management practices; and
23. Uses and potential contamination sources on adjacent property.

The types of owner/operator documents useful in ascertaining the above information may include:

1. Internal memoranda;
2. Aerial photographs;
3. Geologic, hydrogeologic, soil, and topographic maps and reports;
4. Building plans and site plan;
5. Insurance maps;
6. Correspondence with agencies, including responses to state requests for information and EPA Section 104 Requests;
7. Spill reports;
8. Company waste handling and emergency response policies, procedures, and manuals;
9. Company training program records and materials; and
10. Part A and Part B applications under RCRA.

A review of Iowa DNR and EPA files and databases relating to the property and adjoining properties is a mandatory part of any site investigation. In addition, local health departments or the Iowa Department of Labor and Industry may have files on citizen or employee complaints regarding the facility, and local fire departments generally maintain reports of spills necessitating their services. If time permits, the federal Freedom of Information Act[27] may also be utilized although Iowa DNR files are available on request. As a minimum, the following should be reviewed:

1. All permits required and obtained, permit applications pending, and related memoranda and correspondence;
2. Wells in the area and their uses;
3. Any agency violation notices, complaints, or orders;
4. EPA's Comprehensive Environmental Response, Compensation and Liability Information System (CERCLIS), a database of potential or actual hazardous waste sites nationwide;
5. EPA computer list of "potentially responsible parties" at Superfund sites;
6. EPA summary report on Clean Water Act violators;
7. EPA summary report on Clean Air Act violators;
8. Underground storage tank (UST) data bases;
9. Affidavits that may be on file with the county recorder disclosing the use of the property for disposal of wastes;[28]
10. Local zoning and land use ordinances and maps;
11. Citizen or employee complaints lodged with regulatory agencies; and
12. Litigation files, including both administrative and judicial proceedings and notices of citizens' suits.

16.5 Liability of Secured Lenders and Trustees

16.5.1 Lender Liability and Security Interests

CERCLA provides an exclusion from liability for secured parties. Its definition of "owner or operator" excludes "a person, who, without participating in the management of a . . . facility, holds indicia of ownership primarily to protect his security interest in the . . . facility."[29] As noted in section 5.2.5 of this handbook, several cases interpreting this definition have held lenders to be liable if they foreclose on contaminated property or involve themselves in the operations of the

---

27/ 5 U.S.C. § 552.

28/ Iowa Code § 558.69.

29/ 42 U.S.C. § 9601(20)(A).

property. Iowa statutes similarly provide liability limitations for lenders who do not participate in the management of a facility, hold indicia of ownership primarily to protect a security interest, and do not exacerbate a release of hazardous substances.[30] Because the Iowa courts have not yet interpreted this provision, federal case law may provide some guidance on this issue.

In United States v. Mirabile,[31] the court denied a summary judgment motion of a lender that had arranged to place a loan officer on an advisory board that oversaw operations of the borrower after its bankruptcy. However, another lender that foreclosed on the contaminated property but did not actually take title was held to be entitled to summary judgment dismissing it from the action. The court noted that mere financial ability to control waste disposal practices is not in itself sufficient for the imposition of liability, but management of the affairs of the operator was critical in the court's decision.[32]

In United States v. Maryland Bank & Trust Co.,[33] a lender was held liable for cleanup costs even where it did not know of the toxic waste problem at the time it foreclosed on the property and where the problem was caused solely by the previous operator of the site. In United States v. Fleet Factors Corp.[34] a lender which had not foreclosed was nonetheless held liable under CERCLA where it had offered management advice to its borrower and had retained a liquidator that had moved drums containing hazardous waste and engaged in activities which allegedly released friable asbestos. The court indicated that a secured creditor will be liable if its involvement with management of the facility "is sufficiently broad to support

---

30/ Iowa Code §§ 455B.381(7); 455B.392(7) (providing that under certain circumstances lender may be liable for lesser of cleanup costs or postcleanup fair market value of property); see also First Iowa State Bank v. Iowa DNR, 502 N.W.2d 164 (Iowa 1993).

31/ 15 Envtl. L. Rep. (Envtl. L. Inst.) 20,994, at 20,996 (E.D. Pa. Sept. 4, 1985).

32/ 15 Envtl. L. Rep. (Envtl. L. Inst.) 20,994, at 20,997 (E.D. Pa. Sept. 4, 1985).

33/ 632 F. Supp. 573, 578–80 (D. Md. 1986); see also Guidice v. BFG Electroplating & Mfg. Co., 732 F. Supp. 556, 561–63 (W.D. Pa. 1989).

34/ 901 F.2d 1550 (11th Cir. 1990), cert. denied, 111 S. Ct. 752 (1991).

the inference that it could affect hazardous waste disposal decisions if it so chose."[35] As a result of the Fleet Factors decision, lenders became increasingly concerned and confused with regard to what constitutes participation and management sufficient to void the secured creditor exemption. The lending industry continues to press for amendments clarifying CERCLA's secured creditor exemption and, although several bills have been proposed in Congress, none has yet been adopted.

The EPA responded to the Fleet Factors decision by promulgating a rule, effective April 29, 1992, which attempted to clarify the secured creditor exemption under CERCLA, thereby provide guidance for lenders as to the scope of permissible conduct and the conditions under which they may be found liable for the cost of cleaning up contaminated property.[36] The "lender liability rule" was vacated in a February 1994 decision handed down by the U.S. Circuit Court of Appeals for the District of Columbia.[37] The court, which has exclusive jurisdiction to hear challenges to regulations under the federal Superfund law, found that EPA had exceeded its statutory rulemaking authority. Until Congress amends the Superfund statute to clarify lender liability, secured lenders are forced to govern their day–to–day activities by interpreting an array of court rulings that have been neither uniform in rationale nor ultimate conclusion.[38]

Relying on the different statutory authority in the federal Resource Conservation and Recovery Act (RCRA), the EPA has proposed similar lender liability regulations that set forth conditions under which secured creditors may be exempt from corrective action, various technical, and financial responsibility requirements that otherwise apply to underground storage tank (UST) owners and

---

35/ 901 F.2d 1550, 1558 (11th Cir. 1990), cert. denied, 111 S. Ct. 752 (1991).

36/ "Lender Liability Rule Under CERCLA," 40 C.F.R. § 300.1100.

37/ Kelley v. EPA, 15 F.3d 1100 (D.C. Cir. 1994).

38/ United States v. McLamb, 5 F.3d 69 (4th Cir. 1993); Waterville Indus., Inc. v. Finance Auth., 984 F.2d 549 (1st Cir. 1993); Ashland Oil, Inc. v. Sonford Prods. Corp., 810 F. Supp. 1057, 1060 (D. Minn. 1993). But see United States v. Fleet Factors Corp., No. CV687–070 (S.D. Ga. 1993).

operators.[39] Under these proposed UST regulations, which are expected to be finalized in 1995, a security interest holder would not be considered an "owner" for purposes of the corrective action or technical standards so long as the holder does not participate in management or engage in petroleum production, refining, and marketing. Similarly, a security interest holder would not be considered an "operator" for purposes of the corrective action or technical standards so long as the holder prior to foreclosure "is not in control of and does not have responsibility for the daily operation of the UST or UST system." After foreclosure, a security interest holder can avoid treatment as an "operator" provided that the holder within 15 days following foreclosure:

1. empties all of the USTs and UST systems;
2. leaves vent lines open and functioning;
3. caps and secures all other lines; and
4. properly closes the UST system in accordance with EPA requirements.

As a result of these developments, lenders remain concerned with the condition of properties that they are asked to finance. Not only can cleanups impair the financial viability of the borrower, but if the site is contaminated, the lender may not be able to foreclose on it to protect its security interest or may become involved in litigation.[40] Thus, lenders now require and sophisticated buyers willingly agree to undertake a comprehensive site investigation to provide comfort to the lender that, if the buyer should default, the lender could acquire the property by foreclosure without incurring liability for cleanup expenses.

39/ 42 U.S.C. § 6991b(h)(9); 59 Fed. Reg. 30,448 (June 13, 1994); see also Iowa Code § 455B.471(6)(b) (providing for secured creditor protection from UST liability as an "owner" so long as post-foreclosure actions are intended to protect the collateral secured by the interest and the lender seeks to sell or liquidate the secured property rather than holding the property for investment purposes). UST regulation is discussed further in chapter 10 of this handbook.

40/ See First Iowa State Bank v. Iowa DNR, 502 N.W.2d 164 (Iowa 1993) (DNR lacked authority under Iowa Code § 455B.307(1) to compel lender to clean up solid waste already existing on property prior to foreclosure); see also Nischke v. Farmers & Merchants Bank & Trust, 1994 Westlaw 425,177 (Wis. App. Oct. 24, 1994) (regarding UST liability of lender).

Typically, the lender also will require representations and warranties regarding the condition of the site and an indemnification agreement by the buyer, seller, or both. These are of little value, however, where the borrower defaults on the loan for financial reasons since, if it cannot repay the loan, it is unlikely to make good on breach of warranty claims or an indemnification agreement.

16.5.2 Trustee Liability

Although there have been few cases on point, many trustees are concerned that they may incur environmental liabilities merely because they hold legal title to, or operate, real property trust assets. In the preamble to the former EPA lender liability rule, EPA indicated that the rule would not cover trustee and fiduciary concerns, because neither CERCLA's secured creditor exemption, nor any other provision of CERCLA, makes any special provision for trustees. EPA did, however, suggest that a trustee is not personally liable for CERCLA cleanup costs solely because a trust asset is contaminated even though, in most instances, the trust's assets would be available for cleanup of a trust property.[41]

In City of Phoenix v. Garbage Services Co.,[42] the court held that a trustee holding only bare legal title is an owner under CERCLA and thus liable for response costs. Where the trustee is liable as the current owner, the trustee's liability is limited to the extent of the trust assets. Where a trustee is held liable as an owner at the time of the disposal of hazardous substances, but the trustee did not have the power to control the use of the trust property, the trustee's liability is likewise limited to the extent of the trust assets. However, where the trustee had the power to control the use of the trust property, and knowingly allowed the property to be used for the disposal of hazardous waste, the court held the trustee liable under

41/ 57 Fed. Reg. 18,349 (Apr. 29, 1992); see also United States v. Burns, No. C-88-94-L, 1988 Lexis 17,340 (D. N.H. Sept. 12, 1988).

42/ City of Phoenix v. Garbage Services Co., 816 F. Supp. 564 , 568 (D. Ariz. 1993); City of Phoenix v. Garbage Services Co., No. C 89–1709 SC, 1993 U.S. Dist. LEXIS 5970 (D. Ariz., Apr. 6, 1993).

CERCLA to the same extent that it would have been liable if it held the property free of trust. Due to trustees' concerns, amplified by the City of Phoenix case, trustees are also likely to require environmental due diligence prior to taking property into trust and to proceed with caution with regard to existing trust assets that may be contaminated.

16.6 Lease Agreements

Tenants may be liable for contamination of leased premises as "operators" under CERCLA.[43] In addition, tenants may find themselves being sued by their landlords for contamination discovered on the property during their lease period.[44] Defending against a landlord's claim that the contamination did not pre–date the tenant's activities may be difficult after the tenant has conducted its operations on the property over a lengthy period. Similarly, an owner of real estate who leases property to others may incur liability for contamination caused or exacerbated by tenants' activities. Thus, both parties have an interest in ensuring that responsibility for pre–existing contamination is allocated in the lease documents. Traditional leases do not provide sufficient environmental protection to the parties.

Landlords may also wish to obtain higher security deposits, set up escrow accounts to cover future damages, and negotiate provisions which require periodic inspection and environmental assessments to determine if contamination is occurring. In addition, a landlord may seek to restrict the tenant's use of the property for certain purposes, require compliance with environmental laws, and require notification to the landlord of any release of hazardous substances.

On the other hand, tenants may wish to have an environmental assessment performed prior to entering into a lease to protect the tenant from claims of

---

43/ 42 U.S.C. § 9601(20A); see Versatile Metals, Inc. v. Union Corp., 693 F. Supp. 1563, 1571 (E.D. Pa. 1988) (holding lessee liable as operator); see also United States v. South Carolina Recycling & Disposal, Inc., 653 F. Supp. 984, 1003 (D.S.C. 1984) (holding lessee liable as owner).

44/ Caldwell v. Gurley Ref. Co., 755 F.2d 645 (8th Cir. 1985).

contamination occurring during the leasehold period. Tenants may also be wise to perform an assessment at the end of the lease period to demonstrate that the property was uncontaminated when returned to the control of the landlord. In addition, tenants should consider seeking an indemnity from the landlord regarding the presence of any hazardous materials except for those specifically placed on the property by the tenant, and the indemnification agreement should survive the lease termination. Moreover, tenants may wish to obtain representations and warranties about past uses of the property, location of any underground storage tanks, and other potentially harmful conditions such as the presence of asbestos. In properties containing asbestos, the lease should clearly address responsibility for compliance with laws and regulations which exist at the time the lease is signed and which may be enacted during the term of the lease. Finally, the tenant should obtain a default clause allowing it to terminate the lease in the event that a government–mandated or other required cleanup prevents or interferes with the tenant's operation and use of the property.

16.7 Representations and Warranties

Representations and warranties dealing with environmental factors have become commonplace in transactional documents involving transfers of real property. Buyers typically attempt to obtain as many specific representations and warranties in the purchase agreement regarding the property and its past and present uses as possible. Sellers frequently seek to obtain representations from buyers that they have inspected the property; that they have conducted or had an opportunity to conduct an environmental assessment to determine the condition of the property; that the property is free from contamination at the time of the transfer; that the buyer had an opportunity to inspect the seller's files, records, and properties; and that they are entering the transaction on a fully informed basis.

In addition to definitions of basic environmental terms such as "hazardous substance," "release," and "applicable environmental laws," provisions should typically be included in purchase agreements to memorialize the parties' understanding as to the following:

1. Condition of the property;
2. Seller's compliance with applicable laws relating to handling, storage, and disposal of hazardous substances;
3. Seller's noninvolvement with any release;
4. Seller's possession of all permits and approvals necessary for its business and use of the property;
5. Seller's compliance with all approvals and permits;
6. Knowledge of any contamination on the site from any source;
7. Cause of any contamination;
8. Knowledge of any violations, claims, administrative proceedings, or lawsuits relating to hazardous substances on the site;
9. Agreement that Seller will handle and dispose of all hazardous substances in compliance with law prior to closing;
10. No Superliens or other liens due to hazardous waste cleanup have been imposed on the property;
11. Pollution control equipment is in good working condition;
12. Buyer's right of access to and inspection of the property and records;
13. Buyer's right to undertake an environmental assessment of the property, including the taking of soil and groundwater samples.
14. Presence and proper registration of storage tanks and compliance status; and
15. Presence of wells.

16.8 Indemnification and Hold Harmless Agreements

As a practical matter, an indemnification or hold harmless agreement is the principal means of protecting buyers or lenders in real estate transactions involving property with known or suspected contamination. Indemnification provisions should be as specific as possible while allowing for protection from unforeseen liabilities. The form of an indemnification agreement depends on the circumstances relating to the specific site and transaction. Typically, four issues arise in negotiating such an agreement:

1. **Breadth or nature of claims covered.** Any and all claims, damages, and losses arising from the condition of the site? Site investigation costs? Cleanup costs? Claims for personal injuries or property damage by local residents? Natural resources damages? Civil penalties? Buyer's consulting and attorneys' fees to monitor the progress of cleanup and/or administrative proceedings?

2. **Term of the agreement.** *Ad infinitum* or for a specified term of years? Under what circumstances can a cleanup be reopened?

3. **The triggering event.** Can the buyer trigger action by finding contamination itself? Is oral or written notice of a potential claim against any party by any third person sufficient? Is the institution of a formal proceeding by a cognizant governmental agency required?

4. **Extent of losses covered.** All losses without monetary limitation or should losses be capped at a maximum specified amount? Who determines "how clean is clean?"

Other provisions may: (1) require notice to the obligor of any claim; (2) specify who is responsible to undertake and supervise all work; and (3) describe procedures for invoicing and payment of covered costs.

Obviously, the value of an indemnity to a buyer or lender is only as good as the indemnitor's financial strength. The party obtaining an indemnity should keep in mind that the indemnifying party may eventually become insolvent and thus the financial strength of the indemnitor must be ascertained as fully as practicable. If the financial strength is suspect, the acquiring party should consider obtaining a letter of credit, a bond, or escrowing funds sufficient in amount and for a period of time to address environmental problems that may likely arise.

16.9 Effect Of Bankruptcy On Environmental Liability

Environmental laws generally seek to impose environmental cleanup costs on the responsible parties. The Bankruptcy Code, on the other hand, aims to facilitate a "fresh start" by providing an expedient and complete process for a debtor to obtain relief from their indebtedness.[45] Given these divergent objectives, environmental laws and the Bankruptcy Code increasingly have come into conflict, especially in the following five areas:

1. The Bankruptcy Code's automatic stay provision;
2. Discharging a potentially responsible party's (PRP) liability for environmental cleanup;
3. The disallowance of contingent contribution claims;
4. PRP's abandoning contaminated property; and
5. The priority of recovery for environmental cleanup claims.

16.9.1 The Automatic Stay

As soon as a debtor petitions for bankruptcy, the Bankruptcy Code's automatic stay provision bars creditors from pursuing claims that arose prior to the debtor's bankruptcy petition.[46] All judicial, administrative, or other proceedings against the debtor and the enforcement of any judgments obtained before the petition was filed are stayed.

With respect to environmental cleanup claims, however, there is one significant exception to the automatic stay. Section 362(b)(4) of the Bankruptcy Code provides that the stay does not apply to the enforcement of most actions by governmental units seeking to enforce their police or regulatory powers.[47] This exception to the automatic stay has been construed broadly. If the government can show that the cleanup of environmental contamination is necessary to protect

---

45/ The Bankruptcy Code provides individuals, partnerships, and corporations with two distinct means to obtain relief from their indebtedness. In a Chapter 7 bankruptcy, the trustee collects the debtor's property, converts the property to cash, and uses the liquidation proceeds to pay the debtor's creditors. Unlike a Chapter 7 liquidation, a Chapter 11 reorganization contemplates that the debtor will continue to operate its business after the bankruptcy. The Chapter 11 debtor proposes a reorganization plan, which the creditors must approve and the bankruptcy court must confirm, specifying how it will pay its pre-bankruptcy debts. Generally, the type of bankruptcy a PRP pursues does not impact its liability for environmental claims.

46/ 11 U.S.C. § 362(a)(1).

47/ 11 U.S.C. § 362(b)(4)

public health, welfare, morals, and safety, it can avoid the automatic stay and compel a PRP in bankruptcy to complete the environmental cleanup.48/ The police and regulatory power exception does not protect the government's pecuniary interests, however. If a court determines that the government's claim for response costs affects only the government's monetary interests, the claim will be barred by the automatic stay.

16.9.2 Discharge of Environmental Claims

Bankruptcy "discharges" all debts that arose prior to the debtor petitioning for bankruptcy. If a debt is discharged in bankruptcy, creditors are permanently barred from pursuing payment for that debt from the bankrupt debtor.49/

In Ohio v. Kovacs,50/ the U.S. Supreme Court held that certain obligations associated with cleaning up environmental contamination represent a "debt" that may be discharged in bankruptcy. Based on Kovacs, monetary claims (i.e., claims seeking reimbursement for response costs already incurred) brought against a PRP in bankruptcy are discharged. With respect to injunctive orders under CERCLA, the courts have held that liability arising out of an injunctive order becomes a dischargeable debt to the extent the order creates a purely monetary obligation for the bankrupt debtor.51/ However, to the extent an injunctive order does not create an obligation to pay money but "seeks to end or ameliorate pollution," the order does not give rise to a debt which is discharged under the Code.52/

48/ See, e.g., New York v. Exxon Corp., 932 F.2d 1020, 1024 (2d Cir. 1992); United States v. Nicolet, Inc., 857 F.2d 202, 207-10 (3rd Cir. 1988); Penn Terra Ltd. v. Department of Environmental Resources, 733 F.2d 267, 273 (3d Cir. 1984); Hagaman v. New Jersey, 151 B.R. 696, 699 (D.N.J. 1993).

49/ The Code defines "debt" as "liability on a claim." 11 U.S.C. § 101(5)(A), (12).

50/ 469 U.S. 274 (1985).

51/ In re Chateaugay Corp., 944 F.2d 997, 1009 (2nd Cir. 1991).

52/ Chateaugay, 944 F.2d at 1009; see also In re Torwico Elec., Inc., 8 F.3d 146 (3d Cir. 1994).

A controversial "discharge" issue involves the timing of when a bankrupt debtor's environmental liability arises.[53] If liability for environmental contamination arises before the PRP has entered bankruptcy, the PRP's liability for the environmental contamination will be discharged after the debtor has completed the bankruptcy proceedings. Conversely, if the PRP's liability arises after the debtor enters bankruptcy, the debtor will be responsible for the costs of the cleanup following their bankruptcy.

In United States v. Union Scrap Iron & Metal,[54] the district court held that a PRP's liability under CERCLA did not arise until the government has expended funds to clean up pollution caused by the debtor.[55] In other cases, courts have indicated that a CERCLA claim arises for purposes of bankruptcy upon the release or threatened release of hazardous substances into the environment.[56] Employing a third approach, several courts have considered whether the government or PRP seeking contribution had knowledge or could have "reasonably contemplated" the environmental contamination giving rise to the claim prior to the PRP–debtor entering bankruptcy.[57] Under this approach, if the parties asserting claims could have reasonably contemplated the environmental contamination at issue, the

53/ This issue could also arise under the Bankruptcy Code's automatic stay provision since the stay only applies to pre–petition debts.

54/ 123 B.R. 831, 837-38 (D. Minn. 1990).

55/ See also In re Allegheny Int'l, Inc., 126 B.R. 919, 925-26 (W.D. Pa.), aff'd mem., 950 F.2d 721 (3d Cir. 1991). The Minnesota district court has subsequently applied a different approach in determining when a claim arises. See Sylvester Brothers Dev. Co. v. Burlington N. RR, 133 B.R. 648, 652-53 (D. Minn. 1991) (court determined that the parties asserting claims were not aware of environmental contamination prior to PRP-debtor entering bankruptcy; therefore, PRP-debtor was not discharged of post–bankruptcy liability).

56/ In re Chateaugay, 944 F.2d 997 (2nd Cir. 1991), aff'g 112 B.R. 513 (S.D.N.Y. 1990); In re Cottonwood Canyon Land Co., 146 B.R. 992 (Bankr. D. Colo. 1992).

57/ In re Jensen, 995 F.2d 925, 931 (9th Cir. 1993), aff'g 127 B.R. 27 (Bankr. 9th Cir. 1991), rev'g 114 B.R. 700 (Bankr. E.D. Cal. 1990); In re Chicago, Milwaukee, St. Paul & Pacific RR 974 F.2d 775 (7th Cir. 1992); see also In re Chicago, Milwaukee, St. Paul & Pacific R.R. ("Chicago II"), 1993 WL 314078 (7th Cir. Aug. 18, 1993); AM Int'l v. DataCard Corp., 146 B.R. 391, 394 (N.D. Ill. 1992); NCL Corp. v. Lone Star Building Centers, 144 B.R. 170, 177 (S.D. Fla. 1992); In re National Gypsum, 139 B.R. 397, 412 (N.D. Tex. 1992).

PRP–debtor is discharged of post–bankruptcy liability under CERCLA. The "reasonable contemplation" test attempts to balance the conflicting policies of environmental laws and the Bankruptcy Code.[58]

16.9.3 Disallowance of Contingent Environmental Claims

Section 502 of the Bankruptcy Code governs the allowance of claims against a debtor's bankruptcy estate.[59] Most creditor claims are automatically "allowed" under the Bankruptcy Code. Section 502(e) of the Bankruptcy Code, however, which governs the treatment of claims filed by entities that are jointly liable with the debtor, specifically disallows certain claims. Section 502(e) provides:

> (e)(1) [T]he court shall disallow any claim for reimbursement or contribution of an entity that is liable with the debtor on or has secured the claim of a creditor, to the extent that . . .
>
> (B) such claim for reimbursement or contribution is contingent as of the time of allowance or disallowance of such claim for reimbursement or contribution.

Several courts that have considered the allowance of CERCLA contribution claims have found all CERCLA claims disallowed pursuant to Bankruptcy Code § 502(e).[60] However, some courts have held that when a party seeks to recover response costs directly under CERCLA § 107(a)(4)(B), as opposed to recovering costs

---

58/ See Kevin J. Saville, Discharging CERCLA Liability in Bankruptcy: When Does a Claim Arise?, 76 Minn. L. Rev. 327 (1991); see also Shawn F. Sullivan, Discharge of CERCLA Liability in Bankruptcy: The Necessity For a Uniform Position, 17 Harv. Envt'l L. Rev. 445 (1993).

59/ See 11 U.S.C. § 502.

60/ See, e.g., Syntex Corp. v. The Charter Co. (In re The Charter Co.), 862 F.2d 1500, 1503 (11th Cir. 1989); In re Cottonwood Canyon Land Co., 146 B.R. 992, 996-97 (Bankr. D. Colo. 1992); Eagle–Picher Industries, Inc., 144 B.R. 765, 769 (Bankr. S.D. Ohio 1992); In re Bicoastal Corp., 141 B.R. 231, 233-34 (Bankr. M.D. Fla. 1992); In re Kent Holland Die Casting & Plating, Inc., 125 B.R. 493, 501-02 (Bankr. W.D. Mich. 1991); see also In re Hemingway Transp. Inc., 993 F.2d 915, 923-25 (1st Cir. 1993) (suggesting that CERCLA § 107(a)(4)(B) claim would be disallowed unless party asserting claim qualified for CERCLA's "innocent landowner" defense).

owed to or incurred by a government entity under CERCLA § 113(f), that party's CERCLA claims are allowable under Bankruptcy Code § 502(e).[61/]

16.9.4 Abandonment of Property

Under Section 554 of the Bankruptcy Code, "the trustee may abandon any property of the estate that is burdensome to the estate or that is of inconsequential value and benefit to the estate."[62/] Relying on this language, bankruptcy trustees have attempted to avoid environmental liability by abandoning contaminated property that is owned by the debtor.[63/] Addressing this issue, the U.S. Supreme Court in Midlantic National Bank v. New Jersey Department of Environmental Protection[64/] held "that a trustee may not abandon property in contravention of a state statute or regulation that is reasonably designed to protect public health or safety from identified hazards."[65/] Courts have interpreted the Midlantic decision to require resolution of two issues: (1) "whether abandonment would contravene a state law designed to protect the public health and safety;" and (2) "what conditions are necessary to protect the public health and safety."[66/] With respect to the second

---

61/ See In re Allegheny Int'l, Inc., 126 B.R. 919, 923 (W.D. Pa. 1991), aff'd mem., 950 F.2d 721 (3d Cir. 1991); In re Dant & Russell, Inc., 951 F.2d 246 (9th Cir. 1991); In re New York Trap Rock Corp., 153 B.R. 648, 651-52 (Bankr. S.D.N.Y. 1993); Matter of Harvard Inds., Inc., 138 B.R. 10 (Bankr. D. Del. 1992).

62/ 11 U.S.C. § 554.

63/ Abandonment will not eliminate the estate's liability under various environmental liability theories that are not dependent upon current ownership of the property. For example, abandonment will not relieve a generator or operator of liability under CERCLA. See 42 U.S.C. § 9607(e)(1).

64/ 474 U.S. 494 (1985).

65/ 474 U.S. 494 , 506-07 (1985).

66/ See In re Franklin Signal Corporation , 65 B.R. 268, 272–73 (Bankr. D. Minn. 1986). Courts have also generally allowed abandonment when no assets exist in the estate or if the debtor's assets are encumbered. See, e.g., In re Smith Douglas, Inc., 856 F.2d 12 (4th Cir. 1988).

requirement, courts have allowed abandonment when it will not present an immediate danger to the public health.[67]

16.9.5 Priority of Claims

Administrative expense claims are the "actual, necessary costs and expense of preserving the estate."[68] Obtaining administrative expense priority will enable a creditor to maximize its recovery from the bankrupt debtor in that the creditor will have priority status over other unsecured claims.[69] Pre-petition damages under CERCLA caused by the debtor are not entitled to administrative expense priority, but are treated as generally unsecured claims.[70] Most courts, however, have found that post-petition costs expended to remedy pre-petition environmental damage are entitled to administrative expense priority.[71] The rationale for granting post-petition environmental cleanup costs administrative expense priority is that the cleanup of the contaminated property will enhance the value of the property and therefore help preserve the debtor's estate.

16.10 Environmental Considerations in Acquisitions and Mergers

16.10.1 SEC Disclosure Requirements

Regulation S–K governs the filing of forms under the Securities Act of 1933, the Securities Exchange Act of 1934, and the Energy Policy and Conservation Act of

---

67/ In re Anthony Ferrante & Sons, Inc., 119 B.R. 45, 49-50 (D.N.J. 1990); In re FCX, Inc., 96 B.R. 49, 54-55 (Bankr. D.N.C. 1989); In re Oklahoma Refining, 63 B.R. 562, 565 (Bankr. W.D. Okla. 1986).

68/ 11 U.S.C. § 503(b)(1)(A).

69/ 11 U.S.C. § 507(a)(1).

70/ In re Dant & Russell, Inc., 853 F.2d 700, 709 (9th Cir. 1988); In re Great Northern Forest Products, Inc., 135 Bankr. 46, 60 (Bankr. W.D. Mich. 1990).

71/ See, e.g., In re Chateaugay, 944 F.2d 997 (2d Cir. 1991); In re Dant & Russell, Inc., 853 F.2d at 709; In re Great Northern Forest Products, Inc., 135 Bankr. at 61; In re Wall, Tube & Metal Products Co., 831 F.2d 118, 124 (6th Cir. 1987).

1975.[72] Disclosure requirements of legal proceedings expressly include environmental claims. "Administrative or judicial proceedings . . . arising under any federal, state or local provisions . . . for the purpose of protecting the environment" are covered.[73] Disclosure provisions governing the description of the registrant's property designed to "inform investors as to the suitability, adequacy, productive capacity and extent of utilization of the facilities" may also apply to environmental disclosures.[74]

Thus, a party that purchases stock in a company owning contaminated real property may argue that the transaction is subject to Rule 10b–5. In such a case, the buyer may contend that statements by the seller in the stock purchase agreement or financial statements were misleading due to the omission of any reference to a contingent liability for the cleanup of contamination. Assuming the requisite use of instrumentalities of interstate commerce or the mail, the buyer's argument may be colorable under the terms of Rule 10b-5 if the allegedly non-disclosed contingent liability proves to have been material at the time of the sale.

The most difficult issues in such cases will be the highly fact-dependent elements of materiality and scienter. Ultimately, some of these cases may turn on factual issues, including: (1) the certainty of the amount of such contingent liabilities; (2) the magnitude of the cleanup liabilities as compared with the value of all corporate assets; (3) the existence of laws imposing potential remedial liabilities at the time of the stock sale; and (4) the extent to which the selling shareholder can be charged with knowledge of the contamination at the time of the sale.[75]

The SEC issued an interpretive Release on May 18, 1989 to provide guidance to registrants in preparation of management's discussion and analysis reports

---

72/ 17 C.F.R. §§ 229.10 et seq.

73/ 17 C.F.R. § 229.103.5.

74/ 17 C.F.R. § 229.102.1.

75/ See M. Hickok & J. Hoffman, Section 10b–5 Claims for Non–Disclosure of Contingent Cleanup Liabilities, 3 Toxics L. Rep. 315 (Aug. 3, 1988).

(MD&A) pursuant to Item 303 of Regulation S–K. This release does not amend the MD&A requirement, but it adopts an interpretation which modifies many persons' prior understandings about MD&A.[76]

On June 8, 1993, the SEC released Staff Accounting Bulletin (SAB) No. 92, which addresses accounting and disclosures for certain loss contingencies, primarily those relating to environmental costs. The SAB promotes timely recognition of contingent losses and addresses the diversity in practice with regard to accounting and disclosure of such losses.[77]

16.10.2 Special Considerations in Acquisitions

In terms of avoiding environmental liabilities, the buyer is generally best protected with an assets rather than a stock purchase. As discussed in section 7.2.4 of this handbook, even with an assets purchase, exceptions exist to the general rule that the acquiring corporation does not assume the liabilities of the selling corporation.[78] These include: (1) the express assumption of liabilities; (2) the transaction is a de facto merger or consolidation; (3) the successor is a mere

---

76/ See Schneider, MD&A Disclosure, Review of Securities and Commodities Regulation 149, 154 (August 23, 1989); see also J. Archer, T. McMahon, & M. Crough, SEC Reporting of Envtl. Liabilities, 20 Envtl. L. Rep. (Envtl. L. Inst.) 10,105 (Mar. 1990).

77/ Staff Accounting Bulletin No. 92, Securities and Exchange Commission, 17 C.F.R. pt. 211.

78/ Henn & Alexander, Laws of Corporations § 341, at 967 (3rd ed. 1983 & Supp. 1986). The EPA has indicated that another exception recognized in products liability cases, called the "product line" exception, may hold an acquiring corporation liable under CERCLA if it acquires the assets and substantially continues the business of the seller. EPA Memorandum, "Liability of Corporate Shareholders and Successor Corporations for Abandoned Sites under CERCLA," C.M. Price, Ass't Admin. for Enforcement and Compliance Monitoring (June 13, 1984); see also United States v. Mexico Feed & Seed Co., 980 F.2d 478, 487-90 (8th Cir. 1992).

continuation of the predecessor;[79] and (4) the transaction is a fraudulent effort to avoid liabilities.[80]

Other than identifying the liabilities, if any, that will be assumed by the acquiring corporation, the acquirer has an interest in protecting itself which is similar to that of a purchaser of real estate. Thus, warranties and representations and an indemnification agreement are equally applicable to the acquisition situation. In addition, if the seller retains environmental liabilities, the parties must consider the effect of environmental investigation and remediation activities supervised by the seller at a facility which has been transferred to and is controlled by the buyer. The purchase agreement should anticipate the need for access and should provide for cooperation so that interruptions of operations at the facility will be minimized.

79/ See e.g., Ramirez v. Amsted Indus., Inc., 171 N.J. Super. 261, 408 A.2d 818, 823 (N.J. 1979) (successor had same officers, directors, and management as predecessor); New Jersey Dep't of Transportation v. PSC Resources, Inc., 175 N.J. Super. 447, 419 A.2d 1151, 1154 (N.J. 1980).

80/ See e.g. Michigan v. Thomas Solvent Co., 29 Env't Rep. Cas. (BNA) 1119 (W.D. Mich. 1988); United States v. Plastic Electro-Finishing Corp., 313 F. Supp. 330, 331 (E.D.N.Y. 1970).

CHAPTER SEVENTEEN

POLLUTION PREVENTION AND COMPLIANCE AUDITING

17.0 Introduction

Since 1990, the U.S. Environmental Protection Agency (EPA), the Iowa Department of Natural Resources (DNR), and other regulatory agencies around the country have emphasized multi-media pollution prevention measures in addition to traditional media-specific programs like the air, water, and hazardous waste regulatory programs described earlier in this handbook. The focus of these latter programs on waste management and cleanup is shifting to a strong preference for avoiding the generation of toxic and hazardous waste or, alternatively, recycling or treating it. Coupled with the increasing emphasis on criminal prosecutions for environmental violations and the stringent federal sentencing guidelines, auditing compliance with environmental laws has become an important counterpart to pollution prevention programs. This chapter discusses Iowa's pollution prevention program and issues relevant to compliance auditing.

17.1 Toxic Pollution Prevention

17.1.1 Federal Pollution Prevention Act of 1990

In 1990, Congress adopted the federal Pollution Prevention Act which establishes additional reporting requirements for facilities subject to Emergency Planning and Community Right to Know Act (EPCRA) § 313 (discussed in chapter 8 of this handbook).1/ The Pollution Prevention Act declares it to be the national policy of the United States that:

1. Pollution should be prevented or reduced at the source, whenever feasible;

2. Pollution that cannot be prevented should be recycled in an environmentally safe manner, whenever feasible;

3. Pollution that cannot be prevented or recycled should be treated in an environmentally safe manner, whenver feasible; and

---

1/ 42 U.S.C. § 13106.

4. Disposal or other release into the environment should be employed only as a last resort and should be conducted in an environmentally safe manner.2/

Each facility subject to the EPCRA § 313 reporting requirements must report on the quantity of each listed chemical entering the waste stream prior to recycling, the amount of the chemical from the facility that is recycled during each calendar year, the quantity transferred off-site for recycling, the percentage change from the previous year, source reduction practices used with respect to each such chemical at the facility, among other items. This information is submitted on Form R along with the EPCRA reporting data.

17.1.2 Iowa Toxics Pollution Prevention Program

The Iowa Toxics Pollution Prevention Program is designed to prevent toxic pollution at the source and to minimize the transfer of toxic pollutants from one environmental medium to another.3/ The goal of the state is to encourage pollution prevention through the use of pollution prevention techniques in preference to waste management or pollution control, and through coordination and cooperation between federal, state, and local agencies in the development and administration of a pollution prevention strategy.

By its terms, the Iowa statute applies to "toxics users," defined as RCRA large-quantity hazardous waste generators and persons required to file reports pursuant to EPCRA.4/ Toxics users are encouraged to develop a facility-wide multimedia toxics pollution prevention plan. Plans will be reviewed by the DNR for completeness, adequacy, and accuracy and must be kept on file on the premises of the facility. The toxic pollution prevention plan must establish a program which articulates upper

---

2/ 42 U.S.C. § 13101(b).

3/ Iowa Code § 455B.516.

4/ Iowa Code § 455B.516(13).

management support for the plan, identifies toxic substances used by production processes, and assesses approaches to prevent toxic pollution.[5]

The DNR has carried out multi-media pollution prevention activities primarily through the Waste Reduction Assistance Program (WRAP). The WRAP program was statutorily authorized by the Waste Management Assistance Division Act.[6] Despite numerous amendments, this act has retained source reduction or waste elimination as the preferred means of waste management in Iowa.[7]

The WRAP program offers technical assistance in waste reduction or pollution prevention to entities of more than 100 employees or to any RCRA large-quantity hazardous waste generator. This program has used the experience of retired industry professionals to assist more than 100 companies in reducing solid waste, hazardous waste, and air emissions. The WRAP program performs on-site assessments of methods to prevent pollution, offers a pollution prevention workshop, and maintains a clearinghouse function on pollution prevention strategies.

The WRAP program is non-regulatory in that it is administered by personnel separate from the regulatory enforcement activities of the DNR, sometimes in cooperation with the Iowa waste reduction center for the safe and economic management of solid and hazardous waste at the University of Northern Iowa.[8] Assistance information supplied to or obtained by WRAP personnel for the sole purpose of providing assistance to a WRAP program applicant remains confidential. This assistance information cannot be used by an employee or agent of the state in determining whether to initiate an enforcement action or investigation by the state.[9]

5/ Iowa Code § 455B.518.

6/ Iowa Code §§ 455B.480-455B.488.

7/ Iowa Code § 455B.481(3).

8/ Iowa Code § 268.4.

9/ Iowa Code § 455B.484A.

17.2 Environmental, Health, And Safety Compliance Auditing Basics

Compliance auditing has become a key strategy for pollution prevention and to measure compliance with environmental, health, and safety (EHS) requirements. An environmental compliance audit (sometimes called an environmental, health, and safety compliance audit) of a facility examines current processes, equipment, and procedures to determine whether the facility complies with EHS laws. Like a financial audit, which no company would go without, a compliance audit examines operations in light of applicable EHS regulations. Compliance audits generally focus on compliance with prospective laws that address currently generated wastes or on-going activities.

Although most environmental laws do not explicitly mandate compliance auditing, audits commonly are undertaken to accomplish the following:

- Assess whether required permits are in place;
- Evaluate the potential cost of fines and penalties for violations;
- Determine the risk of criminal liability associated with violations;
- Determine the cost of complying with new laws and regulations;
- Discover ways to reduce risk associated with the manner in which processes or operations are conducted;
- Review waste minimization practices; and
- Maximize use of recyclable materials and energy conservation.

An environmental management audit or management systems assessment involves a review of a company's environmental management systems to determine whether the systems insure and preserve compliance and minimize environmental risk. The federal sentencing guidelines discussed below are commonly used as a benchmark to assess the quality of environmental management systems.

17.3 OSHA Process Safety Management Compliance Auditing

As discussed in section 8.8 of this handbook, OSHA regulations mandate compliance auditing in connection with process safety management of highly hazardous chemicals.10/ The process safety management regulations apply to "highly hazardous chemicals" listed in the regulation and certain flammable liquids. Their purpose is to promote safety by enabling the employer and employees involved in operating the process to identify and understand hazards posed by a process involving highly hazardous chemicals.

The process safety management regulations require employee training, process hazard analysis, emergency planning and response, and incident investigation.11/ Employers must certify that they have evaluated compliance with these requirements at least every three years. The compliance audit must be conducted by a person knowledgeable with the relevant process. A report on the audit findings must be developed. Appropriate responses to deficiencies also must be documented. Employers must retain the two most recent compliance audit reports.12/

The OSHA process safety compliance efforts are likely to be conducted in connection with the Clean Air Act risk management planning requirements discussed in section 8.1.4 of this handbook because of their anticipated overlap.13/ In addition to the express auditing requirements in the process safety management regulations, the OSHA Voluntary Protection Program encourages compliance auditing.14/

10/ 29 C.F.R. § 1910.119.

11/ 29 C.F.R. §§ 1910.38(a), 1910.119(n).

12/ 29 C.F.R. § 1910.110(o).

13/ 42 U.S.C. § 7412(r); 58 Fed. Reg. 54,190 (Oct. 20, 1993); 59 Fed. Reg. 4478 (Jan. 31, 1994).

14/ 47 Fed. Reg. 29025 (July 2, 1982); 50 Fed. Reg. 43,804 (Oct. 29, 1985); 51 Fed. Reg. 33,669 (Sept. 22, 1986); 52 Fed. Reg. 7337 (Mar. 10, 1987).

17.4 EPA Environmental Auditing Policy

In 1986, the EPA published its formal policy encouraging the use of environmental audits. In July 1994, the EPA published a restatement of its policies related to environmental auditing.[15] EPA's July 1994 restatement of environmental auditing outlines the following elements of an effective environmental auditing program:

1. Explicit top management support for environmental auditing and commitment to follow up on audit findings;

2. An environmental auditing function independent of audited activities;

3. Adequate team staffing and auditor training;

4. Explicit audit program objectives, scope, resources, and frequency;

5. A process that collects, analyzes, interprets, and documents information sufficient to achieve audit objectives;

6. A process that includes specific procedures to promptly prepare candid, clear, and appropriate written reports on audit results, desired corrective actions, and schedules for implementation; and

7. A process that includes quality assurance procedures to assure the accuracy and thoroughness of environmental audits.

17.5 U.S. Sentencing Guidelines

As noted throughout this handbook, criminal penalties for violations of EHS laws may include probation, fines, imprisonment, or a combination thereof. The penalty ranges set forth in the relevant federal EHS statutes have been more specifically delineated by the federal sentencing guidelines. The U.S. Sentencing Commission adopted sentencing guidelines applicable to federal crimes committed on or after November 1, 1987. These guidelines set limits on the sentencing court's discretion.

15/ 59 Fed. Reg. 38,455 (July 28, 1994).

In short, sentencing guidelines are mathematical formulas by which judges determine mandated penalties for federal offenses. The purposes underlying sentencing guidelines are to reduce sentencing disparities among sentencing courts and to increase the predictability of federal sentences.16/ Chapter Two, Part Q of the sentencing guidelines sets forth the guidelines generally applicable to environmental offenses committed by individuals. An important part of the sentencing guidelines with regard to compliance is the adjustment available when a defendant can show that it had in place "an effective program to prevent and detect violations of law."17/

Commentary following Sentencing Guideline § 8A1.2 defines this term as follows:18/

> (k) An "effective program to prevent and detect violations of law" means a program that has been reasonably designed, implemented, and enforced so that it generally will be effective in preventing and detecting criminal conduct. Failure to prevent or detect the instant offense, by itself, does not mean that the program was not effective. The hallmark of an effective program to prevent and detect violations of law is that the organization exercised due diligence in seeking to prevent and detect criminal conduct by its employees and other agents. Due diligence requires at a minimum that the organization must have taken the following types of steps:
>
> (1) The organization must have established compliance standards and procedures to be followed by its employees and other agents that are reasonably capable of reducing the prospect of criminal conduct.
>
> (2) Specific individual(s) within high-level personnel of the organization must have been assigned overall responsibility to oversee compliance with such standards and procedures.
>
> (3) The organization must have used due care not to delegate substantial discretionary authority to individuals whom the organization knew, or should have known through the exercise of due diligence, had a propensity to engage in illegal activities.

16/ See Sentencing Reform Act of 1984, 18 U.S.C. §§ 3551-3559 & 28 U.S.C. §§ 991-998.

17/ U.S. Sentencing Guidelines §§ 8C2.5(f), 8D1.1(a)(3).

18/ Sentencing Guideline § 8A1.2, comment (k), 56 Fed. Reg. at 22,788 (May 16, 1991).

(4) The organization must have taken steps to communicate effectively its standards and procedures to all employees and other agents, e.g., by requiring participation in training programs or by disseminating publications that explain in a practical manner what is required.

(5) The organization must have taken reasonable steps to achieve compliance with its standards, e.g., by utilizing monitoring and auditing systems reasonably designed to detect criminal conduct by its employees and other agents and by having in place and publicizing a reporting system whereby employees and other agents could report criminal conduct by others within the organization without fear of retribution.

(6) The standards must have been consistently enforced through appropriate disciplinary mechanisms, including, as appropriate, discipline of individuals responsible for the failure to detect an offense. Adequate discipline of individuals responsible for an offense is a necessary component of enforcement; however, the form of discipline that will be appropriate will be case-specific.

(7) After an offense has been detected, the organization must have taken all reasonable steps to respond appropriately to the offense and to prevent further similar offenses -- including any necessary modifications to its program to prevent and detect violations of law.

The precise actions necessary for an effective program to prevent and detect violations of law will depend upon a number of factors [including factors listed in the guidelines] . . .

An organization's failure to incorporate and follow applicable industry practice or the standards called for by any applicable governmental regulation weighs against a finding of an effective program to prevent and detect violations of law.

On November 16, 1993, the Advisory Group on Environmental Offenses proposed that the U.S. Sentencing Commission adopt additional sentencing guidelines applicable to organizations for environmental criminal violations.[19] These draft sentencing guidelines propose that environmental crimes be punished

19/ 58 Fed. Reg. 65,764 (Dec. 16, 1993).

more severely than most other crimes. While these proposed guidelines may be subject to revisions before taking effect, the sentencing guidelines ultimately adopted will likely be based on the proposed organizational sentencing guidelines.

Like the sentencing guidelines' reduction based on an "effective program to prevent and detect violations of law," the proposed organizational sentencing guidelines allow for a reduction based on a showing of a "commitment to environmental compliance." Because these guidelines include auditing requirements, these compliance auditing standards are important in evaluating or negotiating alleged violations. Specifically, the proposed guidelines at § 9D1.1 indicate that the minimum factors demonstrating a commitment to environmental compliance are:

1. Line Management Attention to Compliance;

2. Integration of Environmental Policies, Standards, and Procedures;

3. *Auditing, Monitoring, Reporting, and Tracking Systems.* The organization has designed and implemented, with sufficient authority, personnel and other resources, the systems and programs that are necessary for:

(i) frequent auditing (with appropriate independence from line management) and inspection (including random, and, when necessary, surprise audits and inspections) of its principal operations and all pollution control facilities to assess, in detail, their compliance with all applicable environmental requirements and the organization's internal policies, standards and procedures, as well as internal investigations and implementation of appropriate, follow-up countermeasures with respect to all significant incidents of non-compliance;

(ii) continuous on-site monitoring, by specifically trained compliance personnel and by other means, of key operations and pollution control facilities that are either subject to significant environmental regulation, or where the nature or history of such operations or facilities suggests a significant potential for non-compliance;

(iii) internal reporting (e.g., hotlines), without fear of retribution, of potential non-compliance to those responsible for investigating and correcting such incidents;

(iv) tracking the status of responses to identified compliance issues to enable expeditious, effective and documented resolution of environmental compliance issues by line management; and

(v) redundant, independent checks on the status of compliance, particularly in those operations, facilities or processes where the organization knows, or has reason to believe, that employees or agents may have, in the past, concealed non-compliance through falsification or other means, and in those operations, facilities or processes where the organization reasonably believes such potential exists.

4. Regulatory Expertise, Training, and Evaluation;
5. Incentives for Compliance;
6. Disciplinary Procedures;
7. Continuing Evaluation and Improvement; and
8. Additional Innovative Approaches.

According to the proposed guidelines, an organization that substantially satisfies each of the factors listed in 1 through 7 above, may also endeavor to demonstrate that additional mitigation, up to the allowable levels, is justified due to its implementation of additional programs or components that it can show are effective and important to carrying out its overall commitment to environmental compliance. The organization would have a heavy burden of persuading the court that its additional program or component contributes substantially to achieving the fundamental objectives of environmental compliance represented by the pertinent factor(s) identified in Part D.

17.6 Confidentiality of Compliance Audits

The increasing use of environmental audit programs has raised concerns about the disclosure of audit findings. In some circumstances, the attorney-client privilege may protect environmental audits as confidential communications. Most courts find that communications between attorneys (or even non-lawyers such as consultants retained by the attorney to assist in rendering legal advice) and corporate employees for the purpose of giving legal advice are protected communications.[20] When an environmental compliance audit report is not kept confidential or is not for the purpose of obtaining informed legal advice, the attorney-client privilege generally will not be recognized.[21]

In contrast to the attorney-client privilege, the work-product doctrine provides limited immunity for materials prepared by an attorney in anticipation of litigation.[22] In most circumstances, the work-product doctrine is not an effective method of protecting the confidentiality of an audit because the audit is usually not prepared in anticipation of litigation.[23]

Another basis for protecting the audit from disclosure is the self-evaluation privilege. The privilege has been recognized in a few cases when the information was obtained during self-evaluation, there is a strong public interest in maintaining the self-evaluation information as confidential, and forcing disclosure might eliminate future candid evaluations.[24] Colorado, Indiana, Kentucky, and Oregon have enacted legislation that, with some variations, creates a self-evaluation privilege for audit reports.

20/ Upjohn v. United States, 449 U.S. 383 (1981); see also United States v. Koval, 206 F.2d 918 (2d Cir. 1961) (applying attorney-client privilege for client communications to a non-lawyer employed by client's lawyer); United States v. Chevron, No. 88-6681, 1989 U.S. Dist. Lexis 12,267 (E.D.Pa. Oct. 16, 1989) (primary purpose of environmental consultant's work must be to gain or obtain legal advice).

21/ In re Grand Jury Matter, 147 F.R.D. 82 (E.D. Pa. 1992) (involving documents generated by environmental consultant relating to alleged hazardous waste violations).

22/ Fed. R. Civ. P. 26(b)(3); Iowa R. Civ. P. 122(c).

23/ Diversified Indus., Inc. v. Meredith, 572 F.2d 596, 604 (8th Cir. 1978) (concluding that attorney's investigative file was not prepared in anticipation of litigation); California v. Southern Pac. Transp. Co., No. CV-S-91LKK (E.D. Cal. Oct. 14, 1993) (environmental report prepared for dual purpose of regulatory responsibility and litigation not protected from discovery).

24/ Bredice v. Doctors Hosp., Inc., 50 F.R.D. 249 (D. D.C.), aff'd, 479 F.2d 920 (D.C. Cir. 1973); Webb v. Westinghouse, 81 F.R.D. 431 (E.D. Pa. 1978); see also Olen Prop. Corp. v. Sheldahl, Inc., No. 91-6446-WDK (C.D. Cal. April 12, 1994) (recognizing privilege applicable to environmental audit memorandum).

17.7 Enforcement

Toxic pollution prevention measures under Iowa law have primarily been voluntary. EPCRA and the federal Pollution Prevention Act are backed by civil enforcement authority of $25,000 per day per violation plus criminal enforcement authority set forth in 42 U.S.C. § 11045. The OSHA compliance auditing requirements discussed in this chapter are backed by civil enforcement authority of up to $70,000 per violation plus criminal enforcement authority in 29 U.S.C. § 666.

INDEX

-A-

-B-

-C-

-D-

-E-

-F-

-G-

-H-

-I-

-J-

-L-

-M-

-N-

-O-

-P-

-R-

-S-

-T-

-U-

-V-

-W-

Government Institutes Catalog

For a complete catalog, contact our address below!

	Title	Pub. Date	Price	PC #
	ENVIRONMENTAL			
	CFR Chemical Lists on CD ROM	1996	$110	4080
	Chemical Guide to the Internet	1996	$69	519
	Clean Air Handbook,2nd Edition	1993	$89	343
	Clean Water Handbook,3rd Edition	1997	$89	512
	EH&S Auditing Made Easy	1997	$79	581
■	EMMI-Envl Monitoring Methods Index for Windows-Network	1997	$537	4082
■	EMMI-Envl Monitoring Methods Index for Windows-Single User	1997	$179	4082
	Environmental Audits,7th Edition	1996	$79	525
	Environmental Engineering and Science: An Introduction	1997	$79	548
	Environmental Guide to the Internet,3rd Edition	1997	$59	578
	Environmental Law Handbook,14th Edition	1997	$79	560
	Environmental Regulatory Glossary, 6th Edition	1993	$79	353
	Environmental Statutes	1997	$69	562
■	Environmental Statutes Book/Disk Package	1997	$207	562
■	Environmental Statutes on Disk for Windows-Network	1997	$415	4060
■	Environmental Statutes on Disk for Windows-Single User	1997	$135	4060
	Environmentalism at the Crossroads	1996	$39	570
	ESAs Made Easy	1996	$59	536
■	GI Environmental Database-Network	1995	$447	4073
■	GI Environmental Database-Single User	1995	$149	4073
	Industrial Environmental Management: A Practical Handbook	1996	$79	515
■	IRIS Database-Network	1996	$237	4078
■	IRIS Database-Single User	1996	$495	4078
	ISO 14000: Understanding Environmental Standards	1996	$69	510
	ISO 14001: An Executive Report	1996	$59	551
	Lead Regulation Handbook	1996	$79	518
	Principles of EH&S Management	1995	$69	478
	Property Rights: Understanding Government Takings	1996	$79	554
	RCRA Hazardous Waste Handbook, 11th Edition	1995	$115	503
	TSCA Handbook, 3rd Edition	1997	$95	566
	Wetland Delineation Manual	1987	$59	367
	Wetland Mitigation: Mitigation Banking and Other Strategies	1996	$75	534
	Wetlands: An Introduction to Ecology, the Law and Permitting	1987	$65	467
	SAFETY AND HEALTH			
	Construction Safety Handbook	1996	$79	547
	Cumulative Trauma Disorders	1997	$59	553
	Forklift Safety	1997	$65	559
	Fundamentals of Occupational Safety & Health	1996	$49	539
	Making Sense of OSHA Compliance	1997	$59	535
	Managing Change for Safety and Health Professionals	1997	$59	563
■	OSHA CFRs Made Easy-Network	1997	$387	4084
■	OSHA CFRs Made Easy-Single User	1997	$129	4084
	OSHA Technical Manual, Electronic Edition	1997	$99	4086
	Safety & Health in Agriculture, Forestry and Fisheries	1997	$125	552
	Safety & Health on the Internet	1997	$39	523
	Safety Made Easy	1995	$49	463

We also carry all 50 Titles of the Code of Federal Regulations.
Call for price and availability.

■ Electronic Products

4 Research Place, Suite 200 • Rockville, Maryland 20850-3226
Tel. (301) 921-2355 • FAX (301) 921-0373
E-Mail: giinfo@govinst.com • Internet: http://www.govinst.com

Government Institutes Order Form

4 Research Place • Rockville, MD 20850-3226 • Tel. (301) 921-2355 • Fax (301) 921-0373
Internet: *http://www.govinst.com* • E-mail: *giinfo@govinst.com*

3 EASY WAYS TO ORDER

1. Phone:
301-921-2355
Have your credit card ready when you call

2. **Fax:** 301-921-0373

3. **Mail:** Government Institutes
4 Research Place, Suite 200
Rockville, MD 20850-3226
USA

PAYMENT OPTIONS

❑ Check (*payable to Government Institutes in US dollars*) $_______

❑ Purchase Order (please attach to this order form)

❑ Credit Card: ☐ VISA ☐ MasterCard ☐ AMERICAN EXPRESS

Exp.___/____

Credit Card No. ______________________________

Signature. ______________________________

CUSTOMER INFORMATION

Bill To: (Please attach your Purchase Order)
Name: ______________________________

GI Account# (*7 digits on mailing label*): ____________________

Company/Institution: ______________________________

Address: ______________________________

____

City: ____________________ State/Province: __________

Zip/Postal Code: __________ Country: __________

Telephone: (   ) ______________________________

Fax: (   ) ______________________________

E-mail Address: ______________________________

Ship To:
Name: ______________________________

Title/Position: ______________________________

Company/Institution: ______________________________

Address: ______________________________

____

City: ____________________ State/Province: __________

Zip/Postal Code: __________ Country: __________

Telephone: (   ) ______________________________

Fax: (   ) ______________________________

E-mail Address: ______________________________

ORDER LIST

✔	Qty.	Product Code	Title	Price

❑ **New Edition Automatic Notification System**

Please enroll me in this No Obligation Standing Order Program for the listed above. I want you to notify me of new editions of these books by sending me an invoice. To receive the new edition of the book, I can just pay the invoice. Or if I do not wish to purchase at this time, I can ignore the invoice. In either case, you will continue to notify me of future new editions. I understand that this is a free service and that I am under no obligation to purchase the product.

Subtotal__________
MD Residents add 5% Sales Tax__________
Shipping and Handling__________

Standard Ground UPS Shipping and Handling

Within US:
1-4 items: $6/item
5 or more items: $3/tem

Outside US:
Add $15 for each item (Airmail)
Add $10 for each item (Surface)

Total Payment Enclosed__________